T0191770

Green Energy and Technology

More information about this series at http://www.springer.com/series/8059

Hermann-Josef Wagner · Jyotirmay Mathur

Introduction to Wind Energy Systems

Basics, Technology and Operation

Third Edition

Hermann-Josef Wagner
Energy Systems and Energy Economics
Ruhr-Universität Bochum
Bochum
Germany

Jyotirmay Mathur
Department of Mechanical Engineering
Malaviya National Institute of Technology
Jaipur
India

ISSN 1865-3529 ISSN 1865-3537 (electronic)
Green Energy and Technology
ISBN 978-3-319-88660-2 ISBN 978-3-319-68804-6 (eBook)
https://doi.org/10.1007/978-3-319-68804-6

Softcover reprint of the hardcover 1st edition 2017

Printed on acid-free paper

This Springer imprint is published by Springer Nature
The registered company is Springer International Publishing AG
The registered company address is: Gewerbestrasse 11, 6330 Cham, Switzerland

Preface

In the past few decades, growth in the wind energy sector has been most phenomenal among all the renewable sources of energy. Consensus exists almost worldwide that for ensuring a sustainable future, wind energy can definitely play an important role.

The present book has been written to satisfy the interest of readers on wind energy converters. The authors have tried to strike a balance between a short book chapter and a very detailed book for subject experts. There were three prime reasons behind doing so: first, the field is quite interdisciplinary and requires a simplified presentation for a person from non-parent discipline. The second reason for this short version of a full book is that both authors have seen students and technically oriented people who were searching for this type of book on wind energy. The third reason and motivation for writing this book was to provide some initial information to people who are embarking on a career in the wind industry. It is this group of people that the present book is targeted at.

This book presents the basic concepts of wind energy in its first two chapters. Chapter 3 deals with the physics and mechanics of operation. It describes the conversion of wind energy into rotation of turbine and the critical parameters governing the efficiency of this conversion. Chapter 4 presents an overview of various parts and components of windmills with a blend of basics and recent advancements. Chapter 5 is dedicated to design considerations while selecting appropriate wind turbines for any site. Design options have been presented with their advantages and disadvantages. Chapter 6 is devoted to the utilization and control of operation of wind turbines. In this chapter, various parameters and methods for optimizing the performance are discussed. Chapter 7 presents the economics and financial issues associated with wind energy systems on the example of India and Germany. Chapter 8 explains Life Cycle Assessment of wind energy converters. The book will be closed by an outlook in Chapter 9.

Attempts have been made to include in this book technological advancements up to the middle of the year 2017.

The authors thank Mrs. Laura-Elvira Graziano for updating figures and statistical data and for information about the developments of new technologies of offshore wind converters.

The authors thank especially Mrs. Manuela Koetter for her great help by typing, improving the tables and figures, and by controlling the text. She has supported all three editions of this book by their work and ideas.

The authors wish the readers a happy journey through the interesting field of wind energy.

Bochum, Germany Hermann-Josef Wagner
Jaipur, India Jyotirmay Mathur

Contents

About the Authors

Prof. Dr.-Ing. Hermann-Josef Wagner is Professor for Energy Systems and Energy Economics at the Ruhr-University of Bochum, Germany. He holds a Diploma degree and a Doctorate in Energy Engineering from Technical University in Aachen, Germany. In the past he worked for the German parliament as well as leading scientist for the energy systems analysis in the Research Center Jülich. His first engagement as professor was for the Technical University of Berlin, fellowed by engagement for the Universities Duisburg and Essen. His relevant experiences are on the fields on energy systems analysis, renewable energies such as wind energy and life cycle analysis. He published about 260 articles in international and national journals and books. He is editor of the book series "Energy and Sustainability".

He is member in the International Association for Energy Economics (IAEE) and in the German Association of Engineers (VDI) where he held the chairmanship of the division of Energy and Environment (GEU) with about 24,000 members a lot of years. He was also admitted as member by the German Academy of Sciences Leopoldina and worked as an academy consulter for energy politics.

The German President decorated him for his engagement with the Order of Merit of the Federal Republic of Germany. Contact: email: lee@lee.ruhr-uni-bochum.de.

Prof. Dr.-Ing. Jyotirmay Mathur is Professor of Mechanical Engineering at the Malaviya National Institute of Technology, Jaipur, India. He is also adjunct Professor at Centre for Energy and Environment at his institute. He specializes in the areas of renewable energy systems, energy policy modelling and energy efficiency. He has been member of development of several codes and standards related to mechanical engineering systems in India. He has published more than 70 research papers in highly reputed international journals. Dr. Mathur has been internationally engaged with collaborative projects with researchers in Germany, USA and UK. Contact: email: jyotirmay.mathur@gmail.com and jmathur.mech@mnit.ac.in.

List of Figures

List of Tables

Chapter 1
Wind Energy Today

1.1 Status

Wind energy is one of the oldest sources of energy used by mankind, comparable only to the use of animal force and biomass. Ancient cultures, dating back several thousand years, took advantage of wind energy to propel their sailing vessels. There are references to windmills relating to a Persian millwright in 644 AD and to windmills in Persia in 915 AD. These early wind energy converters were essential for pumping water and grinding cereals. In Europe, wind wheels were introduced around 1200 BC probably as an after-effect of the crusade to the orient. These European windmills were mainly used for grinding, except in the Netherlands where wind wheels supplied the power to pump river water to the land located below sea level. Between 1700 and 1800 AD the art of windmill construction reached its peak. The construction knowledge was relatively high and improved through trial and error. Later on, theories were developed, e.g. those by Euler, providing the tools to introduce new designs and, thus, to substantially improve the efficiency of energy conversion. Many windmills were built and operated in Denmark, England, Germany and the Netherlands during the eighteenth century. In 1750, the Netherlands alone had between 6,000 and 8,000 windmills in operation. The number of windmills in Germany has been estimated at about 18,000 in 1895, 11,400 in 1914, and between 4,000 and 5,000 in 1933.

Around the beginning of the twentieth century, windmills were further improved and the design of a multi-blade farm windmill originated in the USA. By the middle of the century, more than 6 million windmills were in operation in the USA. Worldwide, many of these wind wheels were used to produce mechanical power or as decentralized electricity suppliers on large farms. When the central electricity grid reached every farmhouse at the beginning of the twentieth century, the use of electricity produced by windmills rapidly decreased and the converters were taken out of operation as soon as the next repair job was due. In the nineteen fifties, pioneers like Huetter at the University of Stuttgart, Germany, took up developing

H.-J. Wagner and J. Mathur, *Introduction to Wind Energy Systems*, Green Energy and Technology, https://doi.org/10.1007/978-3-319-68804-6_1

Table 1.1 Use of wind energy world wide

Status of installed wind power		
	Rated capacity 31.12.2016 (GW)	Share worldwide (%)
China	169	35
USA	82	17
Germany	50	10
India	29	6
Spain	23	5
UK	15	3
France	12	3
Canada	12	2
Brazil	10	2
Italy	9	1
Remaining countries	76	16
Total	487	100

and testing modern wind wheels again. Their design is quite different from the previous one; there are only two or three blades with very good aerodynamic parameters, able to rotate at high speed. Due to the high number of rotations, only a small generator is needed to produce electricity. Nevertheless, it was not possible to break even economically selling wind generated electricity in the fifties and sixties of the twentieth century. In the aftermath of the so called oil crisis in the seventies, there was a surge to enforce the development and marketing of wind wheels, especially in the USA, Denmark and Germany. This was based on the understanding that ultimately, additional energy sources emitting less pollution would be necessary. Due to favorable tax regulations in the eighties, about 12,000 wind converters supplying power ranging from 20 kW to about 200 kW were installed in California. In Europe, a lot of tax money was spent on the development of bigger wind converters and on marketing them. Now, at the beginning of 2017, worldwide more than 340,000 windmills of about 487 GW installed power generation capacity are in operation, as shown in Table 1.1. It may be noted that generation of electricity is not essentially in proportion to the installed capacity of wind power. This mismatch indicated towards a very important consideration in selection of site and type of wind mill, capacity utilization factor, that is described at length, later in this book.

In terms of installed capacity, Germany had the leadership until 2007. Then USA had taken over the leadership for couple of years. Over the turn of 2010, China became world leader and continues to lead till date. Out of total 487 GW installed capacity worldwide, China alone had 169 GW by the end of 2016, that is more than the total installed capacity in entire Europe. During initial days, wind turbines of few hundred kilowatt capacity were available in the market, which now has grown up to a capacity of more than 8 MW.

The market introduction of wind energy is proceeding in industrialized countries as well as in development countries like e.g. India. The Indian wind energy sector had an installed capacity of 29 GW at end of 2016. In terms of wind power installed capacity, India is ranked 4th in the World after China, USA and Germany. Still, the total potential of wind energy in the Indian country is far from exhausted.

1.2 Advantages and Disadvantages of Wind Energy Systems

Wind energy offers many advantages, which explains why it is the fastest-growing energy source in the world. Research efforts are aimed at addressing the challenges to increase the use of wind energy.

1.2.1 Advantages

- Wind energy systems are energized by the naturally flowing wind, therefore it can be considered as a clean source of energy. Wind energy does not pollute the air like power plants that rely on combustion of fossil fuels, such as coal or natural gas. Wind turbines do not produce atmospheric emissions that cause acid rain or greenhouse gasses.
- Wind energy is available as a domestic source of energy in many countries worldwide and not confined to only few countries, as in case of oil.
- Wind energy is one of the lowest-priced renewable energy technologies available today.
- Wind turbines can also be built on farms or ranches, thus benefiting the economy in rural areas, where most of the best wind sites are found. Farmers and ranchers can continue to use their land because the wind turbines use only a small fraction of the land. Wind power plant owners make rent payments to the farmer or rancher for the use of the land.

1.2.2 Disadvantages

- Wind power has to compete with conventional power generation sources on a cost basis. Depending on the wind profile at the site, the wind farm may or may not be as cost competitive as a fossil fuel based power plant. Even though the cost of wind power has decreased in the past 10 years, the technology requires a higher initial investment than fossil-fueled solutions for power supply.

- The major challenge to using wind as a source of power is that the wind is intermittent and it does not always blow when electricity is needed. Wind energy cannot be stored; and not all winds can be harnessed to meet the timing of electricity demands. The option of energy storage in battery banks is beyond economically feasible limits for large wind turbines.
- Good wind sites are often located in remote locations, far from cities where the electricity is needed. In developing countries, there is always the extra cost of laying grid for connecting remote wind farms to the supply network.
- Wind resource development may compete with other uses for the land and those alternative uses may be more highly valued than electricity generation.
- Although wind power plants have relatively little impact on the environment compared to other conventional power plants, there is some concern over the noise produced by the rotor blades, and aesthetic (visual) impacts. Most of these problems have been resolved or greatly reduced through technological development or by properly siting wind plants.

1.3 Different Types of Wind Energy Converters: An Overview

Today, various types of wind energy converters are in operation (Fig. 1.1 gives an overview). The most common device is the horizontal axis converter. This converter consists of only a few aerodynamically optimized rotor blades, which for the purpose of regulation can usually be turned about their long axis (Pitch-regulation). Another cheaper way of regulation consists in designing the blades in such a way that the air streaming along the blades will go into turbulence at a certain speed (Stall-Regulation). These converters can deliver power ranging from 10 kW to some MW. The largest converter on the European market has a power of more than 8 MW. The efficiency of this type of wind energy converters in comparison with other types of windmills is very high. Another conventional (older) type of horizontal axis rotor is the multi-blade wind energy converter. It was first built about one hundred years ago. Such windmills have a high starting torque which makes them suitable for driving mechanical water pumps. The number of rotations is low, and the blades are made from simple sheets with an easy geometry. For pumping water, a rotation regulating system is not necessary, but there is a mechanical safety system installed to protect the converter against storm damage. The rotor is turned in the direction of the wind by using a so called wind-sheet in leeward direction. The mechanical stability of such "slow speed converters" is very high; some have had operation periods of more than fifty years. In order to increase the number of rotations, this type of converter had been improved and equipped with aerodynamically more efficient blades facilitating the production of electricity, where the area of a blade is smaller.

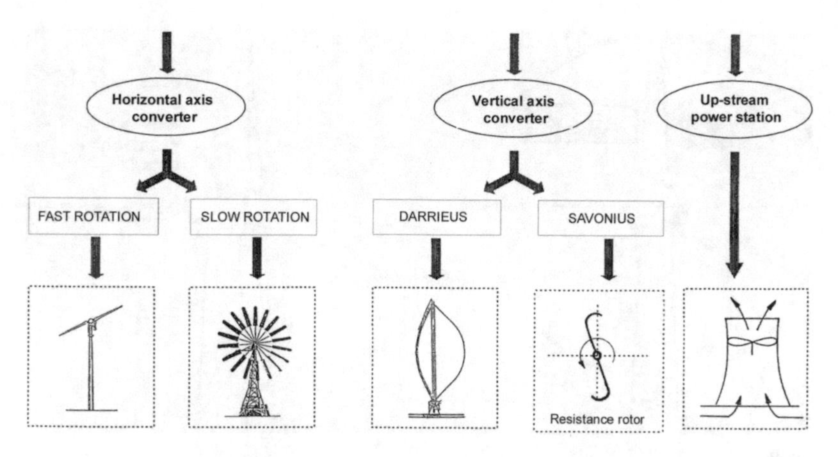

Fig. 1.1 Overview of different types of wind energy converters

A third type of converter is known as DARRIEUS, a vertical axis construction. Their advantage is that they do not depend on the direction of the wind. To start, they need because of their low starting torque the help of a generator working as a motor or the help of a SAVONIUS rotor installed on top of the vertical axis. In the nineteen eighties, a reasonable number of Darrieus-converters were installed in California, but a further expansion into the higher power range and worldwide has not taken place. One reason may be that wind velocity increases significantly with height, making horizontal axis wheels on towers more economical. A modification of the Darrieus rotor is in the form of H-rotor (see also Fig. 5.3); but there are in comparison to other wind converters only a few installations of H-rotors but all of them are below the capacity of 1 MW. Prototype testing was successfully completed for this type but the commercial stage is yet to be seen. The SAVONIUS rotor is used as a measurement device especially for wind velocity; it is used for power production for very small capacities under 100 W.

The last technique to deal with is known as Up-Stream-Power-Station or thermal tower. In principle, it can be regarded as a mix between a wind converter and a solar collector. The top of a narrow, high tower contains a wind wheel on a vertical axis driven by the rising warm air. A solar collector installed around the foot of the tower heats up the air. The design of the collector is simple; a transparent plastic foil is fixed several meters above the ground in a circle around the tower. Therefore, the station needs a lot of space and the tower has to be very high. Such a system has a very poor efficiency, only about one percent. The advantage of such a design is its technical simplicity, which may enable developing countries to construct it by themselves. Worldwide, only one Up-Stream-Power-Station, designed by a German company, has been built so far. Since 1981 it worked satisfactorily at the location of Manzarenas in Spain, but in the year 1988 it was destroyed by bad weather. This station had an electrical power of 50 kW, the tower was about 200 m high and the collector had a diameter of approximately 240 m. Another Up-Stream-Power-

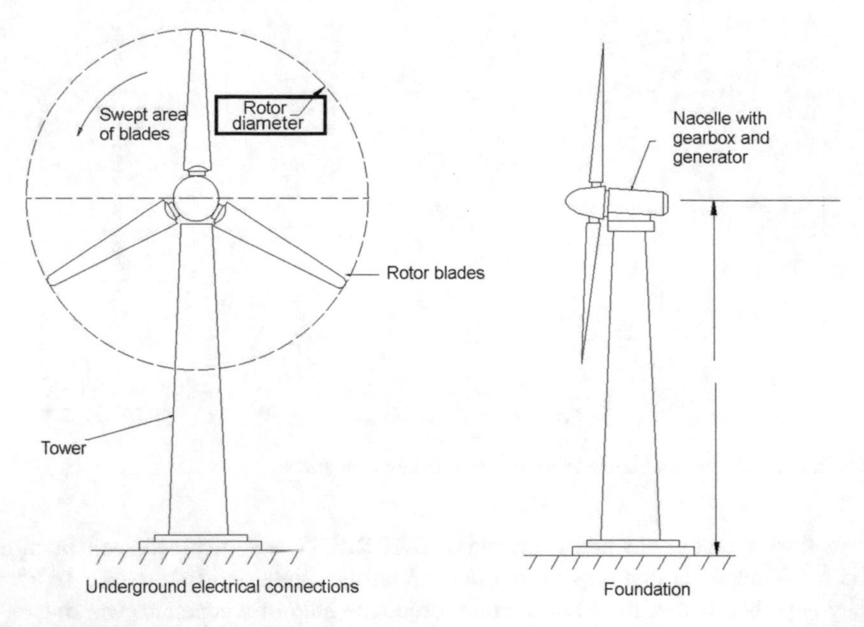

Fig. 1.2 Horizontal axis three blade wind energy converter

Station with an electrical performance of about 200 MW was planned in India. Other feasibility studies have also been conducted in Australia and Namibia but no project of this type has been realized.

Throughout this book, the three terms "wind energy converters", "windmills" and "wind turbines" have been used quite interchangeably. The first term is the technical name of the system, whereas the other two are popularly used terms. Figure 1.2 shows the most popular type of horizontal axis three blade wind energy converter for generating electricity, worldwide. This figure shows the front and side views of a three blade horizontal axis wind energy converter. Details about its major parts and their working are discussed in subsequent chapters.

Literatures

Global Wind Report: Annual Market Update 2016. Global Wind Energy Council, http://www.gwec.net

gwec.net/global-figures/wind-in-numbers 27 June 2017

Chapter 2
Wind: Origin and Local Effects

All renewable energy (except tidal and geothermal power), and even the energy in fossil fuels, ultimately comes from the sun. About 1–2% of the energy coming from the sun is converted into wind energy. This chapter explains the cause of wind flow and factors that affect the flow pattern. Understanding these is necessary for selecting proper locations for wind turbines.

2.1 Origin and Global Availability

The regions around the equator, at 0° latitude, are heated more by the sun than the rest of the globe. Hot air is lighter than cold air; it will rise into the sky until it reaches approximately 10 km altitude and will spread to the North and the South. If the globe did not rotate, the air would simply arrive at the North Pole and the South Pole, sink down, and return to the equator.

Since the globe is rotating, any movement on the Northern hemisphere is diverted to the right, if we look at it from our own position on the ground. (In the Southern hemisphere it is bent to the left.) This apparent bending force is known as the Coriolis force, named after the French mathematician Gustave Gaspard Coriolis.

In the Northern hemisphere, the wind tends to rotate counterclockwise as it approaches a low pressure area. In the Southern hemisphere, the wind rotates clockwise around low pressure areas.

The wind rises from the equator and moves north and south in the higher layers of the atmosphere. Around 30° latitude in both hemispheres the Coriolis force prevents the air from moving much farther. At this latitude, there is a high pressure area, as the air begins to sink down again. As the wind rises from the equator there is a low pressure area close to ground level attracting winds from the North and South. At the Poles, there is high pressure due to the cooling of the air. Table 2.1 shows the prevailing direction of global winds.

H.-J. Wagner and J. Mathur, *Introduction to Wind Energy Systems*, Green Energy and Technology, https://doi.org/10.1007/978-3-319-68804-6_2

Table 2.1 Prevailing global wind directions

Latitude	90–60°N	60–30°N	30–0°N	0–30°S	30–60°S	60–90°S
Direction	NE	SW	NE	SE	NW	SE

The prevailing wind directions are important when siting wind turbines, since one would obviously want to place them in areas with the least obstacles from the prevailing wind directions.

2.2 Different Wind Flows

Winds are very much influenced by the ground surface at altitudes up to 100 m. The winds are slowed down by the earth's surface roughness and obstacles. There may be significant differences between the direction of the global or geostrophic winds because of the earth's rotation (the Coriolis force), and the wind directions near the surface. Close to the surface of the earth, the following effects influence the flow pattern of wind:

(a) **Sea Breezes**

Land masses are heated by the sun more quickly than the sea in the daytime. The air rises, flows out to the sea, and creates a low pressure at ground level which attracts the cool air from the sea. This is called a sea breeze. At nightfall there is often a period of calm when land and sea temperatures are equal. At night the wind blows in the opposite direction. The land breeze at night generally has lower wind speeds, because the temperature difference between land and sea is smaller at night.

The monsoon (specific period of the year when the majority of rainfall occurs) in India and all of South-East Asia is in reality a large-scale form of the sea breeze and land breeze, varying in its direction between seasons, because land masses are heated or cooled more quickly than the sea.

(b) **Mountain Breeze**

Mountain regions display many interesting weather patterns. One example is the valley wind which originates on South-facing slopes (North-facing in the Southern hemisphere). When the slopes and the neighboring air are heated, the density of the air decreases, and the air ascends towards the top following the surface of the slope. At night the wind direction is reversed, and turns into a down-slope wind. If the valley floor is sloped, the air may move down or up the valley, as a canyon wind. Winds flowing down the leeward sides of mountains can be quite powerful.

(c) **The Wind Rose**

It can be noticed that strong winds usually come from a particular direction. To show the information about the distributions of wind speeds, and the frequency of

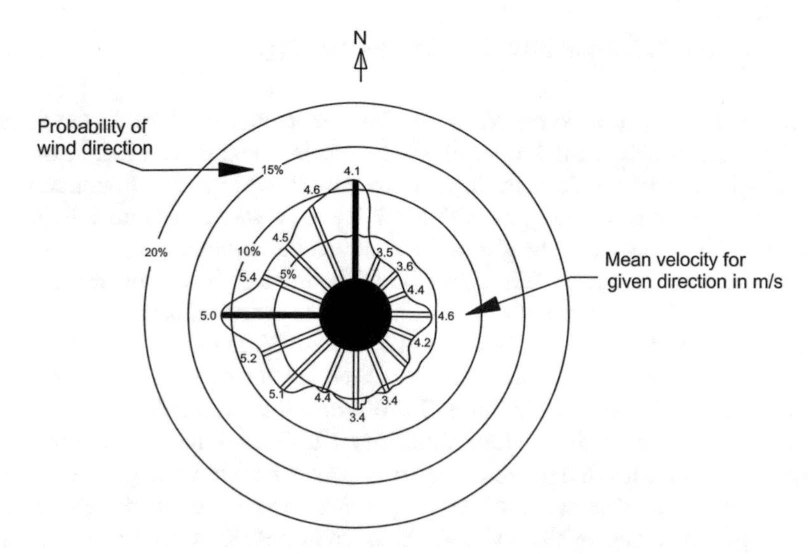

Fig. 2.1 Example of a wind rose (annual all directions average velocity 4.3 m/s)

the varying wind directions, one may draw a so-called wind rose as shown in Fig. 2.1 on the basis of meteorological observations of wind speeds and wind directions.

The wind rose presents a summary of annual wind data. The circular space can e.g. be divided in 16 sectors representing major directions from which wind might come. The number of segments may be more but it makes the diagram difficult to read and interpret. The concentric circles having percent values represent the probability of wind coming from any particular direction. In Fig. 2.1, the polygon in north direction goes slightly ahead of the 10% circle, hence it can be concluded that at this location, the probability of wind coming from the North is roughly 12%. The bars going up to the border of the polygon have a value tag with them. This value represents the mean velocity of wind when it comes from that particular direction. Again, in Fig. 2.1, the bar in the North direction says that when wind comes from the North it has an average velocity of 4.1 m/s. This graph shows that the strongest wind comes from the westward directions.

It should be noted that in popular terms, when we say that some site has 'North wind', this means that wind is coming from the North and not going to the North.

Further to note is that wind patterns may vary from year to year, and the energy content may vary (typically by some 10%) from year to year, so it is best to have observations from several years to calculate a credible average. Planners of large wind parks will usually rely of local measurements, and use long-term meteorological observations from nearby weather stations to adjust their measurements to obtain a reliable long term average.

2.3 Attractive Locations for Wind Energy

In Europe, the area close to the North Sea has the strongest winds. In the extreme North, a wind velocity of the order of 10–11 m/s is available at a height of 50 m. This velocity is very attractive for installation wind machines. The Southern part of Europe also has wind energy potential but it has relatively less wind velocities, in the range of 3–8 m/s at the height of 50 m above the ground.

The National Institute of Wind Energy, India, initially had estimated a 50 GW potential of wind energy in India. This estimate was based on the technologies and economics of wind energy for 50 m and 100 GW for 80 m hub heights around year 2010. High resolution studies have recently been conducted for 100 m hub height, that have revealed total potential of 300 GW, including sites belonging to Rank I: Wasteland, Rank II: Cultivable Land and Rank III: Forest Land. The Southern part of India has been most attractive for wind energy. The wind energy map of India reveals that the maximum wind energy potential lies either in the Southern or Western part of India. In the states of Tamilnadu and Karnataka, there are sites having velocities in the range of 6–8 m/s at a 50 m height above the ground. But in the Western parts of India, Gujarat, close to the Arabian Sea as well as in the desert of Rajasthan, attractive wind energy locations also exist. Recently, due to presence of large coastal line in India and motivated by the success of offshore wind farms in northern Europe, studies have also been initiated to assess wind energy potential using offshore wind farms. All these developments indicate that the potential in other parts of the world, as assessed 10 year ago, also need to be relooked due to development of technology.

The wind data available from various agencies give a fairly good idea about attractive locations for wind energy installations. However, to get exact knowledge about certain locations, measurement of wind availability over several years would be appropriate. The reasons for this are explained in the next section.

2.4 Local Effects on Wind Flow

For the purpose of wind turbines, local wind is perhaps more important, since due to local effects, a site may have very low wind even if it is situated in a predominantly windy area. Major factors that govern local winds are, therefore, described in this section.

2.4.1 Roughness Length and Wind Shear

High above ground level the wind is influenced by the surface of the earth at all. In the lower layers of the atmosphere, however, wind speeds are affected by the friction against the surface of the earth. In the wind industry one distinguishes between the roughness of the terrain, the influence from obstacles, and the influence from the terrain contours, which is also called the orography of the area.

The more pronounced the roughness of the earth's surface, the more the wind will be slowed down. In the wind industry, wind conditions in a landscape are referred through roughness classes or roughness lengths. The term roughness length is the distance above ground level where the wind speed theoretically should be zero. A high roughness class of 3–4 refers to landscapes with many trees and buildings, while a sea surface is in roughness class 0. Concrete runways in airports are in roughness class 0.5.

2.4.2 Wind Speed Variability

The wind speed is always fluctuating, and thus the energy content of the wind is always changing. Exactly how large the variation is depends both on the weather and on local surface conditions and obstacles. Energy output from a wind turbine will vary as the wind varies, although the most rapid variations will to some extent be compensated for by the inertia of the wind turbine rotor. Figure 2.2 shows short term variations in wind.

In most locations around the globe it is more windy during the daytime than at night. This variation is largely due to the fact that temperature differences, e.g. between the sea surface and the land surface, tend to be larger during the day than at night. The wind is also more turbulent and tends to change direction more frequently during the day than at night.

Fig. 2.2 Short term variability of the wind

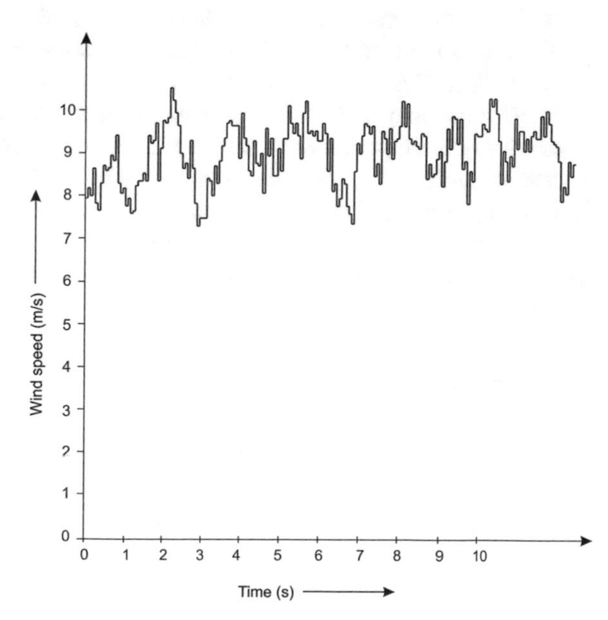

From the point of view of wind turbine owners, it is an advantage that most of the wind energy is produced during the daytime, since electricity consumption is higher than at night. Many power companies pay more for the electricity produced during the peak load hours of the day (when there is a shortage of cheap generating capacity).

At most locations, the wind may not be sufficient for producing power continuously for two to three days and sometimes even for one week.

2.4.3 *Turbulence*

It is normally experienced that hailstorms or thunderstorms in particular, are associated with frequent gusts of wind which both change speed and direction. In areas with a very uneven terrain surface, and behind obstacles such as buildings, a lot of turbulence is similarly created, with very irregular wind flows, often in whirls or vortexes in the neighborhood.

Turbulence decreases the possibility of using the energy in the wind effectively for a wind turbine. It also imposes more tear and wear on the wind turbine, as explained in the section on fatigue loads. Towers for wind turbines are usually made tall enough to avoid turbulence from the wind close to ground level.

2.4.4 *Obstacles to Wind Flow*

Obstacles to the wind such as buildings, trees, rock formations etc. can decrease wind speeds significantly, and they often create turbulence in their neighborhood. It can be seen in Fig. 2.3 that in case of typical wind flows around an obstacle, the turbulent zone may extend to some three times the height of the obstacle. The turbulence is more pronounced behind the obstacle than in front of it. Therefore, it is best to avoid major obstacles close to wind turbines, particularly if they are upwind in the prevailing wind direction, i.e. "in front of" the turbine.

However, every tower of a wind converter will work as such an obstacles and will have influence on the blade when it is coming before or behind the tower.

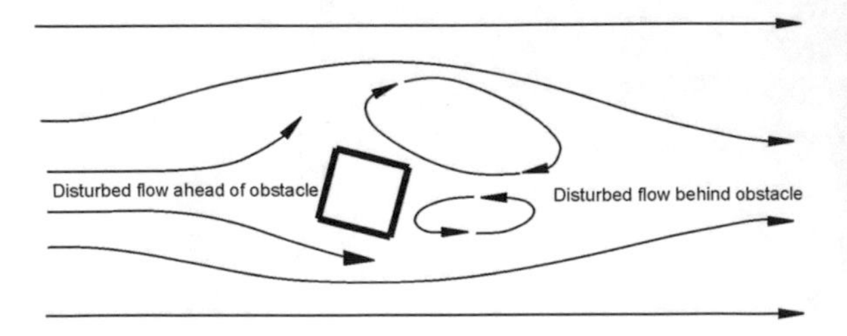

Fig. 2.3 Obstacle in wind flow

Obstacles will decrease the wind speed downstream. The decrease in wind speed depends on the porosity of the obstacle, i.e. how "open" the obstacle is. Porosity is defined as the open area divided by the total area of the object facing the wind. A building is obviously solid, and has no porosity, whereas a fairly open tree may let more than half of the wind through. In case of very dense trees, the porosity is less, say one third. The slowdown effect on the wind from an obstacle increases with the height and length of the obstacle. The effect is obviously more pronounced close to the obstacle, and close to the ground.

(a) **Reduction in Obstacles with Turbine Hub Height**

The higher a turbine is above the top of the obstacle, the less wind shade will be produced. The wind shade, however, may extend to up to five times the height of the obstacle at a certain distance. If the obstacle is taller than half the hub height, the results are more uncertain, because the detailed geometry of the obstacle, (e.g. differing slopes of the roof on buildings) will affect the result.

(b) **Distance Between Obstacle and Turbine**

The distance between the obstacle and the turbine is very important for the shelter effect. In general, the shelter effect will decrease as one moves away from the obstacle, just like a smoke plume becomes diluted as we move away from a smokestack. In terrain with very low roughness (e.g. water surfaces), the effect of obstacles (e.g. an island) may be measurable up to 20 km away from the obstacle. If the turbine is closer to the obstacle than five times the obstacle height, the results will be more uncertain, because they will depend on the exact geometry of the obstacle.

2.4.5 *The Wind Wake and Park Effect*

Since a wind turbine generates electricity from the energy in the wind, the wind leaving the turbine must have a lower energy content than the wind arriving in front of the turbine. There will be a wake effect behind the turbine, i.e. a long trail of wind which is quite turbulent and slowed down, as compared to the wind arriving in front of the turbine. The expression wake is obviously derived from the wake behind a ship. Wind turbines in parks are usually spaced at least three rotor diameters from one another in order to avoid too much turbulence around the turbines downstream.

As a result of the wake effect, each wind turbine will slow down the wind behind it as it pulls energy out of the wind and converts it to electricity. Ideally, therefore turbines should be spaced as far apart as possible in the prevailing wind direction. On the other hand, land use and the cost of connecting wind turbines to the electrical grid would force to space them closer together.

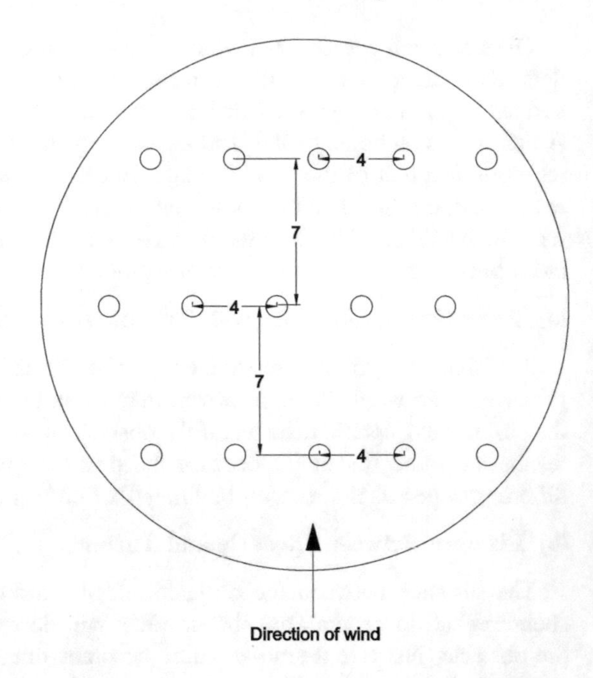

Fig. 2.4 Spacing between turbines in a wind park in terms of rotor diameters (e.g. 4 means four times the rotor diameter)

As a guideline for wind park design turbines in wind parks are usually spaced somewhere between 5 and 9 rotor diameters apart in the prevailing wind direction, and between 3 and 5 diameters apart in the direction perpendicular to the prevailing winds. In Fig. 2.4, three rows of five turbines each are placed in a fairly typical pattern. The turbines (the dots) are placed 7 diameters apart in the prevailing wind direction, and 4 diameters apart in the direction perpendicular to the prevailing winds.

2.4.6 *The Tunnel Effect and Hill Effect*

(a) **The Tunnel Effect**

While walking between tall buildings, or in a narrow mountain pass, it can be noticed that wind velocity increases. The air becomes compressed on the windy side of the buildings or mountains, and its speed increases considerably between the obstacles to the wind. This is known as a "tunnel effect". So, even if the general wind speed in open terrain may be, say, 6 m/s, it can easily reach 9 m/s in a natural "tunnel". Placing a wind turbine in such a tunnel is one clever way of obtaining higher wind speeds than in the surrounding areas. To obtain a good tunnel effect the tunnel should be "softly" embedded in the landscape. In case the hills are very rough and uneven, there may be lots of turbulence in the area, i.e. the wind will be whirling in a lot of different (and rapidly changing) directions. If there is much turbulence it may negate the wind speed advantage completely, and the changing winds may inflict a lot of useless tear and wear on the wind turbine.

(b) **The Hill Effect**

A common way of siting wind turbines is to place them on hills or ridges overlooking the surrounding landscape. In particular, it is always an advantage to have as wide a view as possible in the prevailing wind direction in the area. On hills, one may also experience that wind speeds are higher than in the surrounding area. Once again, this is due to the fact that the wind becomes compressed on the windy side of the hill, and once the air reaches the ridge it can expand again as its soars down into the low pressure area on the lee side of the hill.

2.5 Selecting a Turbine Site

Looking at nature itself is usually an excellent guide to finding a suitable wind turbine site. However, some typical considerations are as follows:

(a) **Wind Conditions**

If there are trees and shrubs in the area, one may get a good clue about the prevailing wind direction. While moving along a rugged coastline, centuries of erosion which have worked in one particular direction can also be noticed. Meteorology data, ideally in terms of a wind rose calculated over 20–25 years is probably the best guide, but these data are rarely collected directly at the site, and there are many reasons to be careful about the use of meteorology data. Meteorologists collect wind data for weather forecasts and aviation, and that information is often used to assess the general wind conditions for wind energy in an area. Unless calculations are made which compensate for the local conditions under which the meteorology measurements were made, it is difficult to estimate wind conditions at a nearby site. In most cases using meteorology data directly will underestimate the true wind energy potential in an area.

If there are already wind turbines in the area, their production results are an excellent guide to local wind conditions. In countries like Denmark, Germany, Spain, and in the Southern part of India, where a large number of turbines are found scattered around the countryside, manufacturers can offer guaranteed production results on the basis of wind calculations made on the site.

(b) **Look for a View**

It is often preferred have as wide and open a view as possible in the prevailing wind direction, and we would like to have as few obstacles and as low a roughness as possible in that same direction. If a rounded hill can be found to place the turbines, one may even get a speed up effect in the bargain.

(c) **Soil Conditions and Transportation Facilities**

Both the feasibility of building foundations for the turbines and road construction to reach the site with heavy trucks must be taken into account with any wind turbine project. Due to the large size of equipment and machinery, these sometimes become bottlenecks for the installation of systems.

To sum up the coverage on wind, it may be said that for taking a macro level decision such as around which city the installation should be planned, the global wind availability and published wind data may be used. However, for an exact siting of wind energy system, a micro level analysis would be necessary due to large variations caused by the local geographical details. Only in case of very high turbines, e.g. 130 m high, where local disturbances have nearly no effect, it may not be necessary to go into micro level detail. The exercise of finding an exact location for installation is termed as 'micro-siting' which requires actual measurement of wind data at the site.

Literature

Wind Energy Resource Map of India, National Institute of Wind Energy, http://www.niwe.res.in

Chapter 3
Physics of Wind Energy

The basic principles of physics on which any wind turbine works are explained in this chapter. These concepts will be helpful in understanding the science and technology behind the operation and control of wind turbines in order to harvest maximum energy from the wind.

3.1 Energy Content in Wind

A wind turbine obtains its power input by converting the force of the wind into torque (turning force) acting on the rotor blades. The amount of energy which the wind transfers to the rotor depends on the density of the air, the rotor area, and the wind speed.

(a) **Density of Air**

The kinetic energy of a moving body is proportional to its mass (or weight). The kinetic energy in the wind thus depends on the density of the air, i.e. its mass per unit of volume.

In other words, the "heavier" the air, the more energy is received by the turbine. At normal atmospheric pressure and at 15 °C, the density of air is 1.225 kg/m^3, which increases to 1.293 kg/m^3 at 0 °C and decreases to 1.164 kg/m^3 at 30 °C. In addition to its dependence upon temperature, the density decreases slightly with increasing humidity. At high altitudes (in mountains), the air pressure is lower, and the air is less dense. It will be shown later in this chapter that energy proportionally changes with a variation in density of air.

(b) **Rotor Area**

When a farmer tells how much land he is farming, he will usually state an area in terms of square meters or hectares or acres. With a wind turbine it is much the same

H.-J. Wagner and J. Mathur, *Introduction to Wind Energy Systems*, Green Energy and Technology, https://doi.org/10.1007/978-3-319-68804-6_3

story, though wind farming is done in a vertical area instead of a horizontal one. The area of the disc covered by the rotor, (and wind speeds, of course), determines how much energy can be harvested in a year.

A typical 1 MW wind turbine has a rotor diameter of 54 m, i.e. a rotor area of some 2,300 m^2. The rotor area determines how much energy a wind turbine is able to harvest from the wind. Since the rotor area increases with the square of the rotor diameter, a turbine which is twice as large will receive 2^2 = four times as much energy.

Figure 3.1 gives an idea of the normal rotor sizes of wind turbines: If the rotor diameter is doubled, one gets an area which is four times larger (two squared). This means that four times as much power output from the rotor will also be obtained.

Rotor diameters may vary somewhat from the figures given above, because many manufacturers optimize their machines to local wind conditions: A larger generator, of course, requires more power (i.e. strong winds) to turn at all. So if one installs a wind turbine in a low wind area, annual output will actually be maximized by using a fairly small generator for a given rotor size (or a larger rotor size for a given generator). The reason why more output is available from a relatively smaller generator in a low wind area is that the turbine will be running more hours during the year.

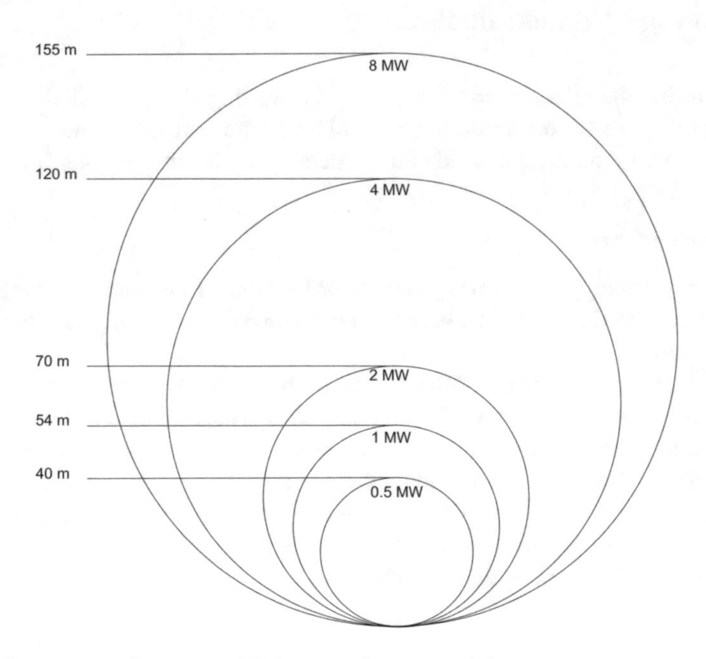

Fig. 3.1 Power output increases with the rotor diameter and the swept rotor area

(c) **Wind Velocity**

Considering an area A and applying a wind velocity v, the change in volume V with respect to the length l and the time t is:

$$\begin{aligned} \Delta V &= A \cdot \Delta l, \\ v &= \frac{\Delta l}{\Delta t} \\ &\Rightarrow \Delta V = A \cdot v \cdot \Delta t. \end{aligned} \tag{3.1}$$

The energy in the wind is in the form of kinetic energy. Kinetic energy is characterized by the equation:

$$E = \frac{1}{2} \cdot m \cdot v^2 \tag{3.2}$$

with m as the mass of wind.

The change in energy is proportional to the change in mass, where

$$m = v \cdot \rho_a \tag{3.3}$$

and ρ_a the specific density of the air. Therefore, substituting for V and m yields

$$E = \frac{1}{2} \cdot A \cdot \rho_a \cdot v^3 \cdot t. \tag{3.4}$$

From the previous equation it can be seen that the energy in the wind is proportional to the cube of the wind speed, v^3. The Power P is defined as

$$P = \frac{E}{t} = \frac{1}{2} \cdot A \cdot \rho_a \cdot v^3. \tag{3.5}$$

Therefore, it is also proportional to v^3. From Fig. 3.2 it can be seen that the power output per m^2 of the rotor blade is not linearly proportional to the wind velocity, as proven in the theory above. This means that it is more profitable to place a wind energy converter in a location with occasional high winds than in a location where there is a constant low wind speed. Measurements at different places show that the distribution of wind velocity over the year can be approximated by a Weibull-equation. This means that at least about 2/3 of the produced electricity will be earned by the upper third of wind velocity. From a mechanical point of view, the power density range increases by one thousand for a variation of wind speed of factor 10, thus producing a construction limit problem. Therefore, wind energy converters are constructed to harness the power from wind speeds in thc upper regions.

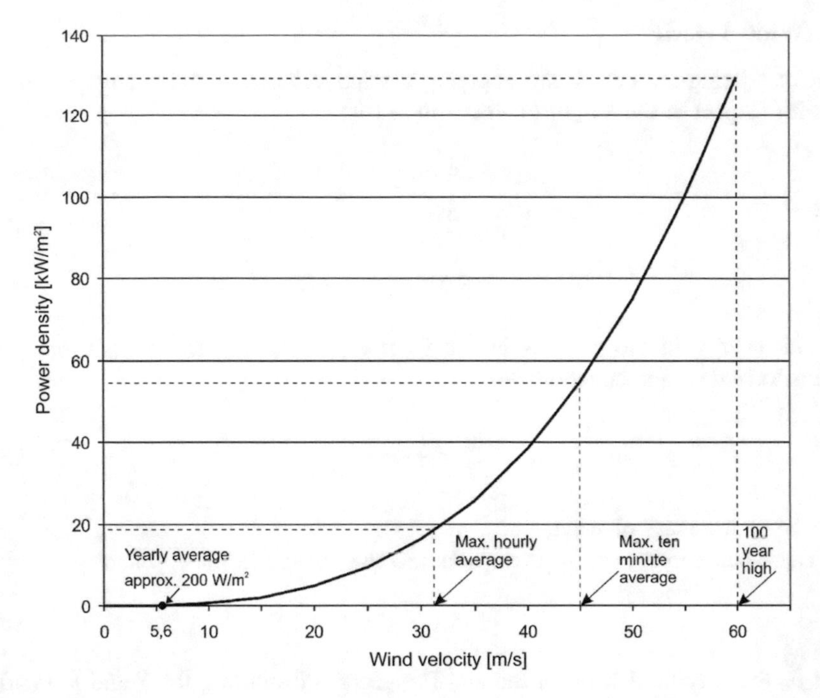

Fig. 3.2 Relationship between wind velocity and power of wind (wind speed for Germany)

3.2 Energy Conversion at the Blade

At the blades, where some kinetic energy of the wind is extracted, the mass of air which passes through the rotor plane gets slowed down. Although kinetic energy is extracted from the airflow at the blades, a sudden step change in velocity is neither possible nor desirable because of the enormous accelerations and forces this would require it to occur. The presence of the turbine causes the approaching air, upstream, gradually to slow down such that when the air arrives at the rotor disc its velocity is already lower than the free-stream wind speed. The stream-tube of air expands as a result of the slowing down in order to accommodate more quantity of air. As the air passes trough the rotor disc, due to aerodynamic action, there is a drop in static pressure such that, on leaving, the air is below the atmospheric pressure level. The air then proceeds downstream with reduced speed and static pressure—this region of the flow is called the wake. Eventually, far downstream, the static pressure and velocity in the wake must return to the level of main stream of air for equilibrium to be achieved.

Figure 3.3 shows the velocities and forces at the profile of a rotating blade. The blade itself moves with an average circumferential velocity u in the plane of the rotor. Rotation of blade provides a relative velocity of air with respect to the blade, which can be considered as additional wind (velocity vector u) working together with the actual wind to decide the rotational force on the blade. The wind flows

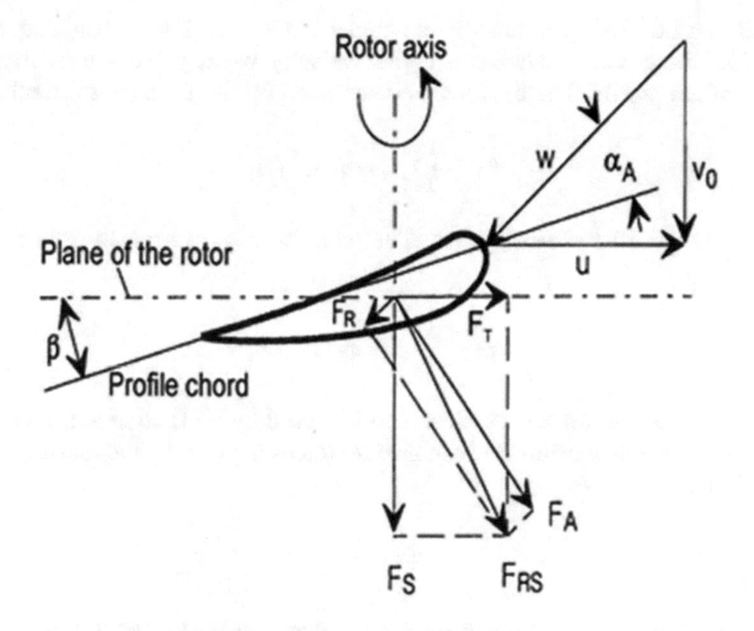

α_A = Angle of attack
β = Pitch Angle
u = Average circumferential velocity
v_0 = Wind velocity in the rotor plane
w = Relative approach velocity
F_R = Drag force - (direction of w)
F_A = Lift force - (vertical direction to w)
F_{RS} = Resultant force
F_T = Tangential component
F_S = Axial component

Fig. 3.3 The velocities and forces acting on a blade

perpendicular to the plane of the rotor, thus creating a resultant velocity vector w. This velocity is then the relative approach or flow velocity of the rotor blade.

The two main forces acting on the rotor blade are the lift force $\boldsymbol{F}_A$ and the drag force $\boldsymbol{F}_R$. The drag force acts parallel to the initial direction of movement and the lift force acts perpendicular to it. The lift force is the greater force in normal operating conditions and arises due to the unequal pressure distribution around an aerofoil profile. The pressure on the upper surface is lower than that on the underside, therefore the air has a higher velocity when passing over the upper surface of the profile. The lift force is determined by the following formula:

$$F_A = \frac{1}{2} \cdot \rho_a \cdot c_A \cdot w^2 \cdot A, \tag{3.6}$$

where c_A is the lift force coefficient. The drag force is determined from a similar formula,

$$F_R = \frac{1}{2} \cdot \rho_a \cdot c_R \cdot w^2 \cdot A, \tag{3.7}$$

where c_R is the drag force coefficient, and is caused by air friction at the surface of the profile. The relationship between the two forces is given by the ratio E_G of their coefficients,

$$E_G = \frac{c_A}{c_R}. \tag{3.8}$$

This relation is at least given by the aerodynamic quality of the blade, by its design and its surface quality. It can be seen from Fig. 3.3 that the resultant force $\boldsymbol{F}_{RS}$ of the lift and drag forces can be divided into two components: the tangentially acting component $\boldsymbol{F}_T$ and the axially acting component $\boldsymbol{F}_S$. It is the force $\boldsymbol{F}_T$ that causes the rotation of the rotor blade and makes power delivery possible.

3.3 Power Coefficients and Principles of Design

The wind turbine rotor must obviously slow down the wind as it captures its kinetic energy and converts it into rotational energy. This means that the wind will be moving slowly after leaving the rotor as compared to its movement before entering the rotor. Farther downstream, the turbulence in the wind will cause the slow wind behind the rotor to mix with the faster moving wind from the surrounding area. The wind shade behind the rotor will therefore gradually diminish as one moves away from the turbine. The factors governing the proportion of velocity ahead and behind the rotor that in turn govern the power output, are discussed in the following sub-sections:

3.3.1 Coefficient of Power c_p *and Betz' Law*

The question of how much of the wind energy can be transferred to the blade as mechanical energy has been answered by the Betz' law.

Betz' law states that only a maximum of 59.25% of the kinetic power in the wind can be converted to mechanical power using a wind turbine, the so called maximum power coefficient or Betz-Number. This number is not higher because the wind on the back side of the rotor must have a high enough velocity to move away and allow more wind through the plane of the rotor.

The relationship between the mechanical power of the rotor blade P_R and the power of wind P in the rotor area is given by the power coefficient c_p:

$$c_p = \frac{P_R}{P}. \tag{3.9}$$

The power coefficient c_p can be interpreted as the efficiency between the rotor blade and the wind. The maximum power coefficient, the above mentioned Betz number, determined by the ratio $v_1/v_2 = 1/3$. Therefore, an ideal turbine will slow down the wind by 2/3 of its original speed.

3.3.2 Tip Speed Ratio

A wind energy converter is classified through the characteristic tip speed ratio λ_S. This is the ratio (as a scalar) of the circumferential velocity of the rotor at the end of the blade (maximum velocity u_e) and the wind velocity v_0 in front of the rotor blade:

$$\lambda_S = \frac{u_e}{v_0}. \tag{3.10}$$

The tip speed ratio has a strong influence on the efficiency of a wind energy converter (Fig. 3.4). When λ_S is small, the circumferential velocity is also small which results in an increase in the angle of attack α_A. When the angle of attack increases past a critical angle, the flow breaks from the profile and becomes turbulent, thus dramatically reducing the lift force. If the tip speed ratio is too large, the lift force will reach its maximum value and decrease afterwards, thus reducing the power efficiency of the converter.

3.3.3 Power Efficiency

The power efficiency of a rotor blade can be determined by investigating the relationship between the power coefficient and the tip speed ratio. Figure 3.5 shows that for every pitch angle β, there is a tip speed ratio λ_S which corresponds to the maximum power coefficient and hence the maximum efficiency. It can be seen that the power efficiency significantly depends on the pitch angle and the tip speed ratio.

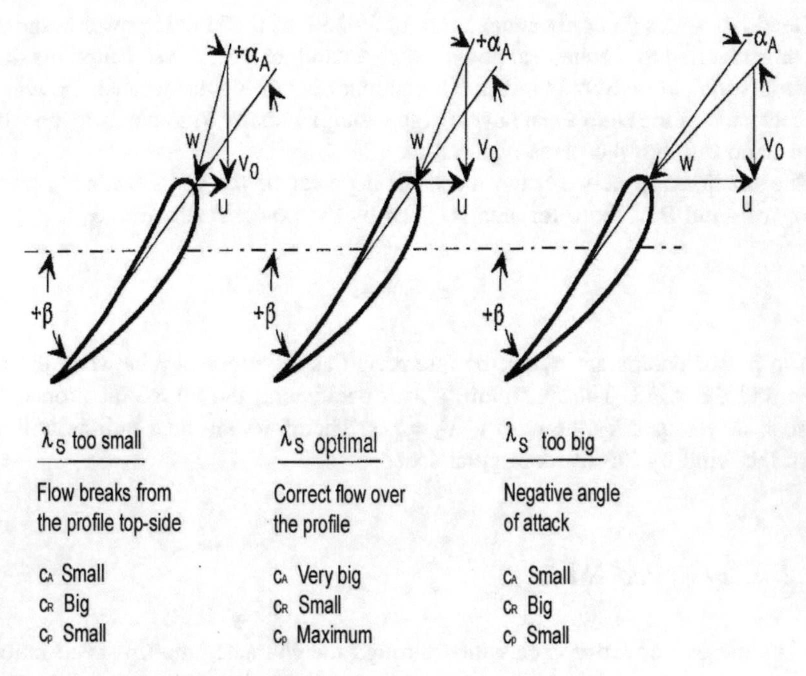

Fig. 3.4 Influence of tip speed ratio λ_S

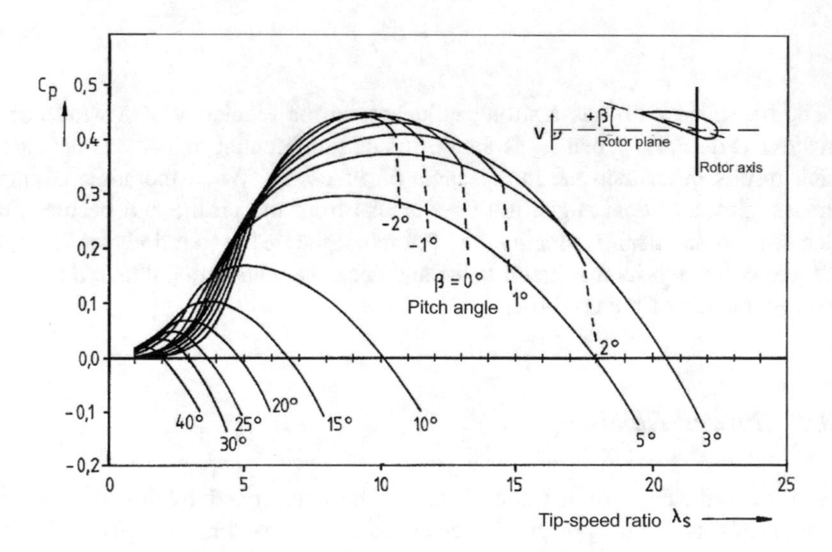

Fig. 3.5 Example of the relationship between the power coefficient and the tip-speed ratio

Therefore, the pitch angle of the blade has to be changed mechanically in respect to the actual tip speed ratio. In case of any pitch angles the power coefficient is negative. These means that the rotor will not turn. It works as a brake.

Disk brake on the main shaft is therefore not necessary in most modern converters by using pitch control.

Besides the power coefficient c_p which can be interpreted as the efficiency between the rotor blade and the wind, there are also energy losses in the mechanical components of the rotor and gear system and in the turbine and generator connection. Therefore, the efficiency can be defined as:

$$\eta = c_p \cdot \eta_m \cdot \eta_{ge}, \tag{3.11}$$

with η_m the mechanical efficiency and η_{ge} the efficiency of the coupled generator and the electrical auxiliary equipment. The efficiency η is also defined by the relationship of the electrical power to the power potential in the wind:

$$\eta = \frac{P_{el}}{\frac{1}{2} \cdot \rho_a \cdot A \cdot v^3}. \tag{3.12}$$

3.3.4 *Principles of Design*

The issues discussed above can be summed up and related to the design of a wind energy converter through the following principles:

1. A high aerofoil form ratio c_A/c_R leads a high tip speed ratio and therefore a large power coefficient c_p.

   → Modern converters with a good aerodynamic profile rotate quickly.

2. Simple profiles with smaller profile form ratios have a small tip speed ratio. Therefore, the area of the rotor radius that is occupied by blades must be increased in order to increase the power coefficient.

→ Slow rotating converters have poor aerodynamic profiles and a high number of blades.

3. The profile form ratio and the tip speed ratio have a considerably greater influence on the power coefficient than the number of blades.

→ For modern converters with a good aerodynamic profile, the number of blades is not so important for a large power coefficient c_p.

3.4 Wind Variations

3.4.1 Wind Shear with Height

Assuming that the wind is blowing at 10 m/s at a height of 100 m, Fig. 3.6 shows how wind speeds vary in agricultural land with some houses and sheltering hedgerows with some 500 m intervals.

The fact that the wind profile shown in Fig. 3.6 is twisted towards a lower speed, as one moves closer to ground level, is usually called wind shear. Wind shear may also be important when designing wind turbines. It may be noticed that for a wind height of 50 m, the wind velocity is 9 m/s whereas for a 100 m height it is 10 m/s in our example case. With the help of the formula for power of wind discussed earlier in this chapter, it can be calculated that due to the dependence of power with cube of wind speed, this increase of 1 m/s corresponds to about 30% difference in the power available with wind. It is also important to note that if a wind turbine with a hub height of 75 m and a rotor diameter of 50 m is considered, one can notice that the wind is blowing at 10 m/s when the tip of the blade is in its uppermost position (100 m height), and 9 m/s when the tip is in the bottom position (50 m height). This means that the forces acting on the rotor blade when it is in its top position are far larger than when it is in its bottom position.

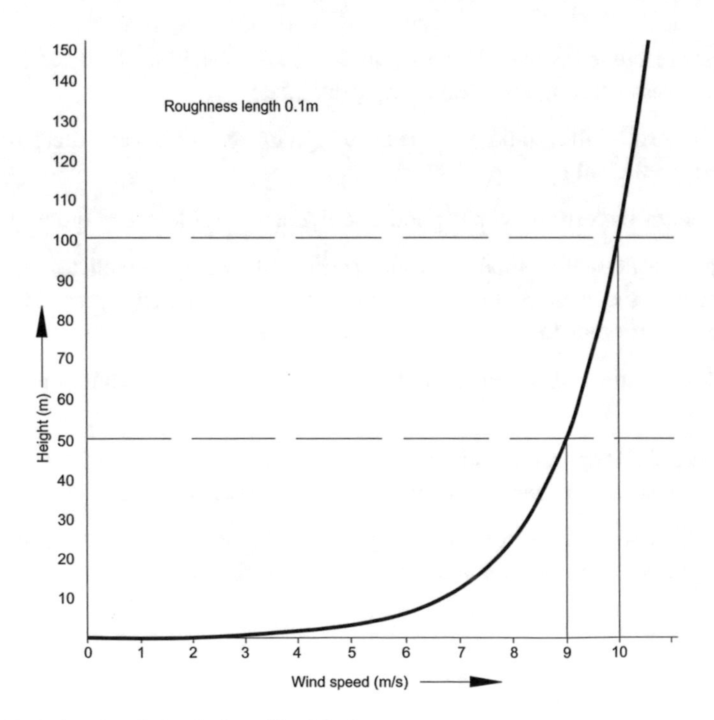

Fig. 3.6 Variation in wind velocity with altitude

Table 3.1 Roughness class and roughness lengths

Roughness class	Roughness length (m)	Energy index (%)	Landscape type
0	0.0002	100	Water surface
0.5	0.0024	73	Completely open terrain with smooth surface, e.g. airport runways, mowed grass
1	0.03	52	Open agriculture land without fencing and very scattered buildings, softly rounded hills
2	0.1	39	Agriculture land with some houses and 8 m tall sheltering hedgerows at a distance of approx. 500 m
3	0.4	24	Villages, small towns
4	1.6	13	Very large cities with tall buildings and skyscrapers

The wind speed at a certain height above ground level is given by:

$$v = v_{ref} \ln(z/z_0)/\ln(z_{ref}/z_0)$$

where

v wind speed at height z above ground level,
v_{ref} reference speed, i.e. a wind speed at height z_{ref},
ln(…) is the natural logarithm function,
z height above ground level for the desired velocity v,
z_{ref} reference height, i.e. the height at which wind speed is v_{ref},
z_0 roughness length in the current wind direction.

A list of all major types of roughness classes and their typical roughness lengths have been given in Table 3.1.

The above example is assuming that the wind is blowing at 9 m/s at 50 m height and the wind speed at 100 m height is to be calculated. If the roughness length is 0.1 m, then:

v_{ref} = 9 m/s
z = 100
z_0 = 0.1
z_{ref} = 50 hence,
v = 9 ln(100/0.1)/ln(50/0.1) = 10 m/s

Average wind speeds are often available from meteorological observations measured at a height of 10 m. Hub heights of modern 1,000–3,000 kW wind turbines are usually 80–130 m and more. Using the above mentioned approach, one may calculate average wind speeds at different heights and roughness classes. It is to be noted that the results are not strictly valid if there are obstacles close to the wind turbine (or the point of meteorological measurement) at or above the specified hub height. It should be rather noted that there may be inverse wind shear on hilltops because of the hill effect, i.e. the wind speed may actually decline with increasing height during a certain height interval above the hilltop. A careful study of wind velocity and profile is therefore recommended before arriving at any conclusions about site.

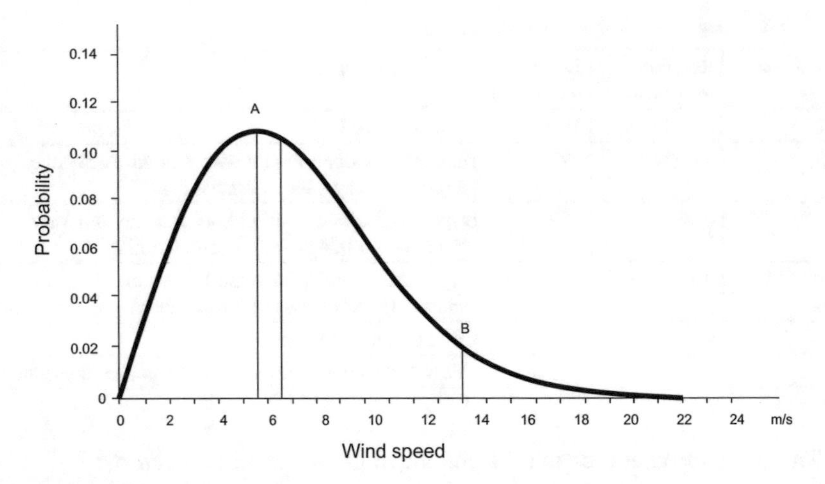

Fig. 3.7 Weibull distribution plot between wind velocity and probability

3.4.2 Influence of Weibull Distribution

It is very important to understand the variation of wind speeds within the studied/ measured time periods since it governs the energy output significantly. If one measures the wind speeds throughout a year, it can be noticed that in most areas strong gale force winds are rare, while moderate and fresh winds are quite common. The impact of such variation on energy output is described in this section.

The wind variation for a typical site is usually described using the so-called Weibull distribution, as shown in Fig. 3.7. This particular site has a mean wind speed of 7 m/s, and the shape of the curve is determined by a so called shape parameter of 2.

It can be realized that the shown graph is a probability density distribution. The area under the curve is always exactly 1, since the probability that the wind will be blowing at some wind speed must be 100%.

Half of the area is to the left of the vertical black line at 6.6 m/s, and half is on its right hand side. The 6.6 m/s is called the median of the distribution shown above. From site-to-site, this median value would be differing. For this site, it means that half the time it will be blowing less than 6.6 m/s, the other half it will be blowing faster than 6.6 m/s. As it can be seen, the distribution of wind speeds is skewed, i.e. it is not symmetrical. Sometimes there will be very high wind speeds, but they are very rare. Wind speeds of 5.5 m/s, on the other hand, are the most common ones. The statistical distribution of wind speeds varies from place to place around the globe, depending upon local climate conditions, the landscape, and its surface.

The Weibull distribution may vary from site to site, both in its shape, and in its median value. There are two parameters that govern the shape of the Weibull distribution curve, namely, the scale parameter and the shape parameter. A higher value of the scale parameter means the distribution is spread over a wider range and the probabilistic average wind velocity has a higher value. A higher value of shape parameter (between 2 and 3) means the distribution is more skewed towards higher wind velocities, if the shape parameter is between 1 and 2, it means that the distribution is skewed towards lower velocities, indicating a higher probability of lower wind velocities. Of course, both these parameters influence the peak distribution curve, but one has major influence on the average value and the other primarily influences the skew ness of the curve. For an exact feeling of curve, both should be considered together.

If the shape parameter is exactly 2, the distribution is known as a Rayleigh distribution. Wind turbine manufacturers often give standard performance figures for their machines using the Rayleigh distribution.

The reason why care is to be taken in connection with wind speeds is their energy content. The wind power varies with the cube of the wind speed. Design/selection of a wind turbine as per mathematical average wind speed may lead to an improper utilization of wind potential at the site. In Fig. 3.7, two conditions marked A and B, one corresponding to the peak velocity i.e. 5.5 m/s which has a probability of 11%, and the other the velocity of 14 m/s having a probability of 2% can be compared. Due to the dependence of power with cube of velocity, at these two velocities 5.5 m/s and 14 m/s, the difference between their cubes, i.e. between 166 and 2.744, is huge. Which means that at the high speed of 14 m/s, more than 16 times the power is available but only for 2% duration, whereas for a large duration of 11% only a small amount of power is available. If the wind turbine is selected as per 14 m/s wind velocity, it would remain much underutilized since the probability of the presence of smaller wind velocities is much higher. However, these figures are to be seen with respect to the energy delivered by the wind turbine and the time/day/month of delivery. Some countries such as Germany, where a feed-in tariff system exists, only the amount of power delivered is important, whereas, at some places the time of delivery may also be important. This analysis that affects the selection of a machine very much, can be done through the power curve of the machine being selected, which is discussed in Chap. 6.

Literatures

Bansal NK, Kleemann M, Meliss M (1990) Renewable energy sources and conversion technology. Tata McGraw-Hill Publishing Company Ltd., New Delhi. ISBN 0-07-4600023-0

Burton T, Sharpe D, Jenkins N, Bossanyi E (2001) Wind energy handbook, Wiley Canada, ISBN 0471489972

Landolt-Börnstein (2006) Energy technologies, Subvolume C, Renewable energy, pp 233–241. Springer Verlag S, Berlin. ISBN 3-540-42962-X

Chapter 4
Components of a Wind Energy Converter

A wind energy converter has the following major components:

1. Rotor blades
2. Gearbox
3. Generator
4. Tower
5. Miscellaneous parts

All of these parts, their major design features and variations available on the market are discussed in this chapter. Besides these major components, there are other necessary parts, such as grid connection. Options for grid connection are covered in Chap. 6.

4.1 Rotor Blades

The basic principles behind the working of wings of airplanes and blades of wind turbines are quite common. However, since the wind turbines actually work in a very different environment with changing wind speeds and changing wind directions, there are special considerations that are not important in the design of airplane wings.

Figure 4.1 shows a photo of a rotor blade for a 5 MW wind converter. Figure 4.2 defines the four sides of a blade.

Due to the aerodynamic profile of blade, as discussed in Chap. 3, low pressure is created on the upper surface of a blade. This creates the lift, i.e. the force pulling upwards, the same principle that enables the plane to stay in the air. In case of the rotor blade of a windmill, the lift is perpendicular to the direction of the wind.

Choosing profiles for rotor blades involves a number of compromises including reliable pitch control and stall characteristics, and the profile's ability to perform well even if there is some dirt on the surface. In countries where rains occur during

H.-J. Wagner and J. Mathur, *Introduction to Wind Energy Systems*, Green Energy and Technology, https://doi.org/10.1007/978-3-319-68804-6_4

Fig. 4.1 Rotor blade of a 5 MW offshore wind turbine (*photo* © DOTI 2009/alpha ventus)

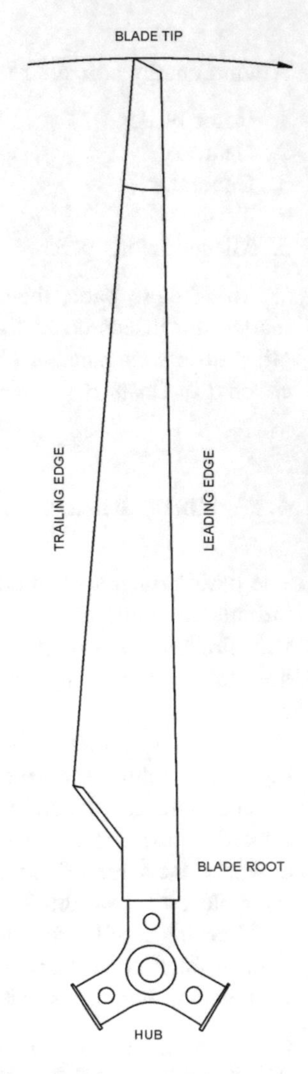

Fig. 4.2 Four sides of a rotor blade of a wind turbine

the whole year, the dirt gets cleaned by the rain water, but in relatively dry countries like India, dirt may become a problem. Even in Germany, when there is only light rain, it may not be sufficient to whip-off the dirt from the blades. Under these circumstances, the blades will have to be cleaned and possibly even polished again after 4–6 years, to ensure a smooth aerodynamic effect.

Most modern rotor blades on large wind turbines are made of glass fibre reinforced plastics, (GRP), i.e. glass fibre reinforced polyester or epoxy. Using carbon fibre or aramid (Kevlar) as reinforcing material is another possibility, usually such blades become necessary in large turbines in the range of 3–8 MW capacity. Wood, wood-epoxy, or wood-fibre-epoxy composites have not penetrated the market for rotor blades. Steel and aluminum alloys have problems of weight and metal fatigue respectively.

Large rotor blades typically consist of a metal walled circular root section with metal bushings with bolts or studs for mounting the blade to the hub. This root section is integrated into a continuously tapered longitudinal beam, the spar, which provides the stiffness and strength required to carry the wind load and the weight of the blade. Around the spar, the two aerodynamic-shaped shells, the suction side and the pressure side, form an optimized aerodynamic wing. The outer shells of the blade meet at the leading edge and the trailing edge. Inserted webs take up the torsional twist of the blade and help stabilize the blade against bending, shear loads and global buckling.

4.2 Gearboxes

The power from the rotation of the wind turbine rotor is transferred to the generator through the power train, i.e. through the main shaft, the gearbox and the high speed shaft. It is not appropriate to just drive the generator directly with the power from the main shaft unless the turbine is equipped with multi-pole generators that are discussed in the next chapter.

If an ordinary generator with two poles (which makes one pair of poles) was used without a gearbox, and the machine was directly connected to a 50 Hz or 60 Hz AC (alternating current) three phase grid, it would become necessary to turn the rotor at 50 revolutions per second i.e. 3,000 revolutions per minute (rpm). With a 50 m rotor diameter that would imply a tip speed of the rotor of far more than twice the speed of sound, a condition which cannot be accepted. Usually, a tip-speed of more than 100 m/s is not accepted across the industry. Another possibility of avoiding a gearbox is to build a slow-moving AC generator with a lot of poles. If a generator was to be connected directly to the grid, this would require a 300 pole generator to arrive at a reasonable rotational speed of 20 rpm.

With a gearbox, slowly rotating, high torque power obtained from the wind turbine rotor can be converted into high speed power, which would be required for the generator. The gearbox in a wind turbine usually does not “change gears”. It normally has a single gear ratio between the rotation of the rotor and the generator. In some machines, the gear ratio is in the range of 30–200.

A typical issue for a gearbox is cooling and heating of the lubricant. In cold climatic conditions like in Germany, in winters, the lubricating oil freezes due to extreme low temperatures. Special heating arrangements are made for keeping the oil sufficiently warm. The situation in hot climates, like in India, is reverse. Arrangements are to be made to prevent the lubricating oil from becoming hot because viscosity of oil decreases with increase in its temperature and therefore at high temperature oil does not serves its purpose of lubricating very well.

A recent development in the field of gearboxes for modern machines is the development of hydrodynamic gearboxes that are actually a combination of a conventional mechanical gearbox and a torque changer. There is a variable speed drive inside and a constant drive outside. The main advantage of this type is continuously variable input but constant output speed adjustment over a wide range. Prototypes have been successfully tested and this technology is very close to the commercialization stage.

4.3 Generators

The wind turbine generator converts mechanical energy to electrical energy. Wind turbine generators are a bit unusual, compared to other generating units ordinarily attached to the electrical grid. One reason is that the generator has to work with a power source (the wind turbine rotor) which supplies very fluctuating mechanical power (torque).

Generating Voltage

In large wind turbines the voltage generated by the turbine is usually around 690 V three-phase alternating current (AC). The current is subsequently sent through a transformer next to the wind turbine (or inside the tower) to raise the voltage to somewhere between 10,000 and 30,000 volts, depending on the standard in the local electrical grid. Manufacturers will supply both 50 Hz wind turbine models (for the electrical grids in most of the world) and 60 Hz models (for the electrical grid e.g. in America).

Generator Cooling System

Generators need cooling while they work. On most turbines this is accomplished by encapsulating the generator in a duct, using a large fan for cooling by air, but a few manufacturers use hydraulically cooled generators. Hydraulically cooled generators may be built more compactly, but they require a heat exchanger (radiator) in the nacelle to get rid of the heat from the liquid cooling system.

Generator Rotational Speed

The speed of a generator which is directly connected to a grid is constant, and dictated by the frequency of the grid. The relationship between the rotational speed of generator and frequency is governed by the formula:

$$\text{Rotation per second} = \frac{\text{Frequency per second (50 Hz or 60 Hz)}}{2 * \text{number of poles}}. \tag{4.1}$$

The number of poles means the number of coils in the stator of the generator which the electric power is generated in. Two poles make one pole pair. The number of poles will be decided by the construction of generator. Because the fact that grid is operating with three phases, the generator must also produce three phase alternating current. Every pole pair consists for construction of six magnets.

When the number of pole pairs is doubled in the stator of a synchronous generator, the rotation will be reduced to half. The variations between rotation per minute and pole pairs are given in Table 4.1.

Often wind turbines that are with gearboxes, use generators with two or three pole pairs. The reasons for using these relatively high-speed generators are savings on size and cost. The maximum force (torque) a generator can handle depends on the rotor volume. The other type of wind turbines that do not use a gearbox but use multipole generators (explained in the next chapter), have a large number of poles.

The number of poles is kept less than 300 (say 84) due to weight considerations. As a result, the frequency of electricity generated is less than 50 Hz, which is transformed into 50 Hz before supply to the grid electronically.

The term “synchronous generator speed“ (f/p, the ratio of frequency and number of pole pairs) thus refers to the speed of the generator when it is running synchronously with the grid frequency. It applies to synchronous generators. However, in the case of asynchronous (induction) generators it is equivalent to the idle speed of the generator. This means that the generator rotation must be a little bit higher than the synchronous speed e.g. in case of 2 pole pairs, 1,515 rpm for a synchronous speed of 1,500 rpm. The advantages and disadvantages of synchronous and asynchronous generators are further discussed in Chap. 5.

Wind turbines which use synchronous generators use electromagnets normally in the rotor which are excited by direct current from the electrical grid. Since the

Table 4.1 Synchronous generator speeds in records per minute (rpm)

Pole number	No. of pole pairs	50 Hz	60 Hz
2	1	3,000	3,600
4	2	1,500	1,800
6	3	1,000	1,200
8	4	750	900
10	5	600	720
12	6	500	600
300	150	20	–
360	180	–	20

grid supplies alternating current, they first have to convert alternating current to direct current before sending it into the coil windings around the electromagnets in the rotor. The rotor electromagnets are connected to the current by using brushes and slip rings on the axle (shaft) of the generator.

Permanent magnet synchronous generators are tested and are coming in the market. Their advantage is to have a low weight.

The key component of the basic version of an asynchronous generator is the squirrel cage rotor or simple cage rotor.

For understanding the operation of an asynchronous generator in basic version, consider manual cranking of this rotor around at exactly the synchronous speed of the generator, e.g. 1,500 rpm, for the 2-pole pair synchronous generator. Since the magnetic field rotates at exactly the same speed as the rotor, there would be no induction phenomena in the rotor, and it would not interact with the stator. The stator is connected to the grid and its field is rotating recording the grid frequency and pole numbers with 1,500 rpm. If the speed of rotor is increased above 1,500 rpm, in that case the rotor moves faster than the rotating magnetic field from the stator, in that condition, once again the stator induces a strong current in the rotor. The harder the rotor is cranked, the more power will be transferred as an electromagnetic force to the stator. The current in the rotor will create an electromagnetic field which generates electricity in the stator, which is fed into the electrical grid.

It is the rotor that makes the asynchronous generator different from the synchronous generator. The rotor consists of a number of copper or aluminum bars which are connected electrically by copper or aluminum end rings. The rotor is placed in the middle of the stator which is directly connected to the three phases of the electrical grid. The clever thing about the cage rotor is that it adapts itself to the number of poles in the stator automatically. The same rotor can, therefore, be used with a wide variety of pole numbers.

Two Speed, Pole Changing Generators

Some manufacturers fit their turbines with two generators, a small one for periods of low winds, and a large one for periods of high winds. A more common design on newer machines is pole changing generators, i.e. generators which (depending on how their stator magnets are connected) may run with a different number of pole pairs, and thus a different rotational speed.

Whether it is worthwhile to use a switching of pole pairs in the generator depends on the local wind speed distribution, and the extra cost of the pole pairs changing generator compared to the price the turbine operator gets for the electricity. However, due to the problems related to a drop in grid voltage, the practice of switching the poles has been discontinued in Europe and many other countries.

4.4 Towers

The tower of the wind turbine carries the nacelle and the rotor. In principle, the tower needs to be as tall as possible, because the wind speed increases with height. How-ever, the height is optimized by analyzing the cost of the increase in tower height and the gain in energy output due to increased wind velocity at a greater height. Towers for large wind turbines may be either tubular steel towers, lattice towers, or concrete towers. Guyed pole towers are only used for small wind turbines.

(a) **Tubular Steel Towers**

Most large wind turbines are delivered with tubular steel towers, which are manufactured in sections of 20–30 m with flanges at either end, and bolted together on the site. The towers are conical (i.e. with their diameter increasing towards the base) in order to increase their strength and to save materials at the same time.

(b) **Lattice Towers**

Lattice towers are manufactured using welded steel profiles. The basic advantage of lattice towers is cost, since a lattice tower requires only half as much material as a freely standing tubular tower with a similar stiffness.

The advantage of the low cost of lattice towers exists only in those countries where the labor cost is cheap e.g. in India. In countries like Germany, where the labor costs are high, the advantage of reduced material cost gets balanced by increased labor costs for assembling the tower. Another disadvantage of lattice towers is their visual appearance, (although that issue is clearly debatable). Be that as it may, for aesthetic reasons lattice towers have almost disappeared from being used for large, modern wind turbines. These towers are used only in small machines i.e. below the MW size.

(c) **Concrete towers**

Recently, with a further increase in the height of towers for more than 100 m hub height, due to the increased cost of steel in towers, manufacturers have started to bring concrete towers in place of steel towers on the market. In future, therefore, there may be an additional type of tower on the wind energy market.

(d) **Guyed Pole Towers**

Many small wind turbines are built with narrow pole towers supported by guy wires. The advantage lies in saving weight and, consequently, cost. The disadvantage lies in a difficult access around the towers which makes them less suitable in farm areas. Such towers are used only in kW size machines.

(e) **Hybrid Tower**

Some towers are built by using different combinations of the techniques mentioned above. It may be a hybrid between a concrete tower (down) and steel (upward) or between a lattice tower and a guyed tower.

Generally, it is an advantage to have a tall tower in areas with a high terrain roughness, since the wind speeds increase with height above the ground, as discussed in previous chapters. Lattice towers and guyed pole towers, due to a reduced surface area as compared to tubular steel tower, have the advantage of giving less wind shade than a massive tower. However, the turbulence created by each element of a lattice structure adversely affects the energy output.

4.5 Miscellaneous Components

Large wind turbines are equipped with a number of safety devices to ensure safe operation during their lifetime. A brief description of such devices is given below:

(a) **Yaw Mechanism**

Almost all horizontal axis wind turbines use forced yawing, i.e. they use a mechanism which uses electric motors and gearboxes to keep the rotor plane perpendicular to the direction of the wind. Almost all manufacturers of upwind machines prefer to apply brakes to the yaw mechanism whenever it is unused. The yaw mechanism is activated by the electronic controller which checks the position of the wind vane on the turbine several times per minute, whenever the turbine is running.

Cables carry the current from the wind turbine generator down through the tower. Occasionally, a wind turbine may look as if it had gone berserk, yawing continuously in one direction for about any revolutions. The cables, however, will become more and more twisted if the turbine by accident keeps yawing in the same direction for a long time. The wind turbine is, therefore, equipped with a cable twist counter, which tells the controller that it is time to untwist the cables. The turbine is also equipped with a pull switch which gets activated if the cables become too twisted.

Besides the role of tracking the wind direction, the yaw mechanism plays another very important role by connecting the tower with the nacelle. Due to the operation of gears, it needs regular lubrication for smooth operation.

(b) **Brakes**

Braking action may be required for several reasons. There are several types of braking a turbine rotor: aerodynamic brakes, electro brakes and mechanical brakes. In case of aerodynamic braking, the blade is turned in such a direction that the lift effect which causes rotation, does not appear. This concept is explained in the section of 'stall control' in Chap. 6. In case of electro-magnetic braking, energy produced by the generator of a wind turbine is dumped into a resistor bank, thereby converting it into heat.

Another type of braking is conventional mechanical braking for which disc brakes are provided in the nacelle. A mechanical drum brake or disk brake is also used to hold the turbine at rest for maintenance. Such brakes are also applied after blade furling and electro braking have reduced the turbine speed, as the mechanical brakes would wear quickly if used to stop the turbine from full speed.

In large wind turbines, normally there is a combination of at least two brakes, most turbines use aerodynamic brakes together with mechanical braking or even also with an electro braking system.

(c) **Over-Speed Protection**

In addition to braking explained above, it is essential that wind turbines stop automatically in case of malfunction of a critical component, for example, if the generator overheats or is disconnected from the electrical grid and the rotor will start accelerating rapidly within a matter of seconds. In such a case it is essential to have an over-speed protection system. Most companies provide wind turbines with two independent fail-safe brake mechanisms to stop the turbine.

(d) **Vibration measurement**

Excessive vibrations may be one of the most dangerous causes for failure of a wind turbine as well as any other component. Vibrations cause fatigue loading of components causing early failure. An additional side effect of vibration is the generation of noise. One of the classical, and most simple safety devices in a wind turbine is the vibration sensor. It simply consists of a ball resting on a ring. The ball is connected to a switch through a chain. If the turbine starts shaking, the ball will fall off the ring and switch the turbine off. Modern turbines have permanently installed vibration monitoring systems which in the case of excessive vibrations decide on the operation of the machine with the help of microprocessor.

(e) **Anemometer**

Before actuating the controlling actions, measurement of wind speed and direction is necessary in every wind turbine. It is usually done using a cup anemometer. The cup anemometer has a vertical axis and three cups which capture the wind. The number of revolutions per minute is registered electronically. Normally, the anemometer is fitted with a wind vane to detect the wind direction. Other anemometer types include ultrasonic or laser anemometers which detect the phase shifting of sound or coherent light reflected from the air molecules. Hot wire anemometers detect the wind speed through minute temperature differences between wires placed in the wind and in the wind shade (the leeward side). The advantagc of non-mechanical anemometers may be that they are less sensitive to icing in cold countries and jamming due to dust in hot locations.

(f) **Lubrication System**

Since the wind turbine has moving parts, lubrication is always a major concern. There are two types of lubrication systems. With the first type, there is a central lubrication pump which sends lubricant to all bearings through tubes laid for this purpose. The other system has a pressurized container of lubricant for each bearing which directly lubricates the bearings.

(g) **Foundation**

Foundation plays a very important role in stabilizing the wind turbine. Due to the large height, heavy weight at the nacelle and large rotor area which faces wind forces, the role of a foundation becomes very important. In case of on-shore installations, the type of foundation is governed by the nature of soil. On the other hand, in case of off-shore installations, where design of foundations is even more serious issue, type is governed by the depth of water in sea. Its designs tends to rely on technology used by the oil and gas industry in some cases. There are different major types of foundations used in case of off-shore wind turbines. The first is the mono pile foundation (see Fig. 4.3) that consists of a steel pile which is driven approximately 10–12 m into the seabed. The second one is the Tripod foundation (Fig. 4.4). The piles on each end are typically driven about 10 m into the seabed, depending on soil conditions. The third is a so called jacked. It is having a lattice tower also, fixed by piles driven in the seabed. The fourth is a so called tripile solution (Fig. 4.5). With the tripile solution the advantages of the simpler monopole construction and transportation can be used, but the necessary load distribution on tops must designed carefully, concerning consequences of possible motions. The next one is the Gravity foundation (also Fig. 4.5). It consists of a large base constructed from concrete which rests through gravity and its own weight on the seabed. The turbine is dependent on gravity to remain erect. Sometimes gravity foundations are further attached with piles into the sea-bed. Monopiles and gravity foundation are usually preferred only in locations where water depth is less than about 20 m. Another type of foundation is the suction bucket, which could be used in off-shore locations having soft seabed. The suction bucket is a steel cylinder with a lid on top which induces a vacuum below the lid by pumping water outside the cylinder (Fig. 4.6).

(h) **Floating Platforms**

Latest among all types of foundations for off-shore wind turbines, have started using a floating platform on which the wind tower is installed. The platform floats a little bit under the water surface and is fixed by guy-wires on sea ground. Such platforms for floating wind turbines should be installed at water depths of 50–500 m similar to the oil and natural gas platforms. Technically it is difficult and expensive to install fixed wind turbines from a sea depth of about 50 m as compared to installing it using a floating platform. With the help of floating offshore wind turbines, the potential of wind energy utilization increases since larger

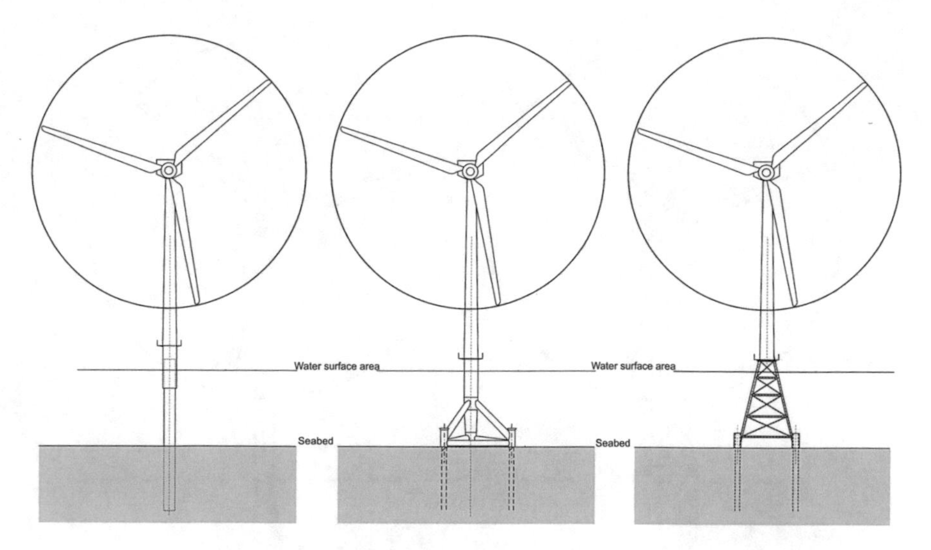

Fig. 4.3 Three foundations for offshore wind turbines. From left to right Monopile, Tripod, Jacked

Fig. 4.4 Tripod foundation for the windpark alpha ventus in Germany (*photo* © Grosse Böckmann)

capacity machines can be used at same locations. The coasts of Japan and the USA fall steeply and are experiencing the floating type of plant very suitable. The platforms currently in use are the Tension-Leg Platform (TLP), the semi-submersible platform and the spar buoy (Fig. 4.7).

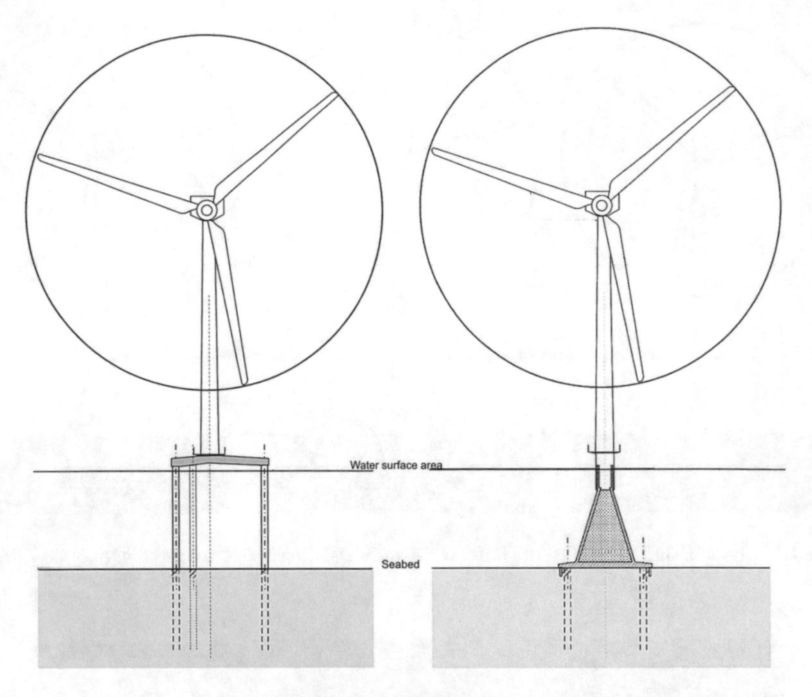

Fig. 4.5 Tripile and Gravity foundation for an offshore wind turbines

Fig. 4.6 Foot of an offshore-suction-bucket (*photo* © Youtube)

The TLP consists of half-dive buoys, which are vertically anchored by rope at the sea-bed. This design offers advantageous of low weight of the TLP and its excellent stability. The semi-submersible platforms consist of a steel frame with vertical cylinders, which are responsible for the buoyancy and are called the buoys. This platform is also attached to the seabed by means of an anchorage. The semi-submersible platforms offer an advantage in their flexible use, in shallow as well as in large water depths. The spar buoy has a simple construction, since the hollow cylinder serves both as a tower and as a platform. It is made relatively heavier in the lower area for ensuring stability and then anchored to the sea-bed.

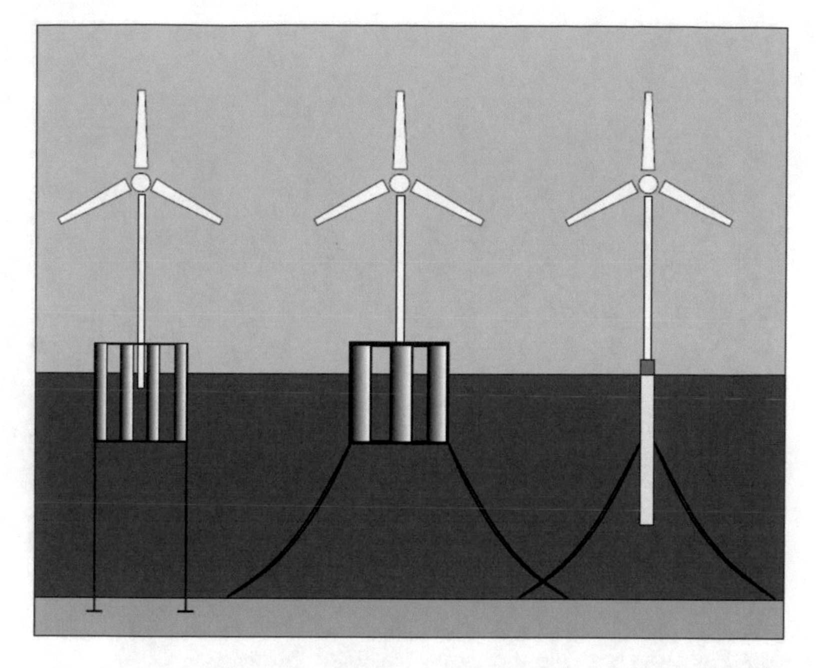

Fig. 4.7 Floating offshore wind turbine platforms. From left to right TLP, Semi-hub, Spar-buoy

Europe is at the forefront of utilizing off-shore wind energy with many floating offshore projects. For example, after a successful trial period of the Hywind pilot project in Norway in 2009, a prototype of floating wind turbine "WindFloat" with an electric power of 2 MW was installed close to the coast of Portugal in the year 2011 (Fig. 4.8). It is standing on three floating cylinders. The cylinders are capable of taking-in different quantities of water for always keeping and stabilizing the wind turbine vertically. Five wind turbines of the 6 MW class was installed in the "Hywind Scotland" project in the year 2017. The location is 25 km far away the coast at Aberdeenshire and the water depth is 90–110 m. Since the Fukushima event in 2011, Japan has been conducting research on various platforms through pilot installation of 14 MW. Other countries, e.g. India, have also started working on pilot installation for offshore wind energy plants. In the next few years, the installation floating offshore wind turbines, worldwide, is likely to expand by many folds.

(i) **Other Parts**

There are many sensors used in the nacelle, e.g. electronic thermometers which check the oil temperature in the gearbox and the temperature of the generator. There are other important measurements that play a major role in connection and operation of a wind turbine with grid: voltage-frequency measurement, phase-angle measurement, revolution measurement, electric power supply measurement. There are self controlling systems with a telecommunication link to a common service point. All of these make the wind turbine more efficient and available for a

Fig. 4.8 Prototype of floating offshore wind turbine "Windfloat" in Portugal (*photo* © https://commons.wikimedia.org/wiki/File:Agucadoura_WindFloat_Prototype.jpg

maximum time for the economic generation of power. The components related to grid connection are discussed in Chap. 6.

j) **Certification**

For improving the quality of windmills different countries have specified their own requirements and methods for testing of wind energy converters, e.g. Germany, Denmark, China, India, Netherlands, Japan.

It may also be noted here that requirements and guidelines for certification and testing of small size wind turbines are significantly different than those for large turbines. This is due to the reason of difference in size, weight, control, power connection, and safety requirements.

Chapter 5
Design Considerations

While finalizing the specifications of wind turbines for installation at any site, a lot of technological options are found to be available. These options along with a discussion on their suitability in various situations are presented in this chapter.

5.1 Rotor Area of Turbines

The question of small or large turbines pertains to the decision related to the area of rotor. A larger rotor with compatible generator would be called a large turbine.

Reasons for Choosing Large Turbines

1. There are economies of scale in wind turbines, i.e. larger machines are usually able to deliver electricity at a lower cost than smaller machines. The reason is that the cost of foundations, road building, electrical grid connection, plus a number of components in the turbine (the electronic control system etc.), are somewhat independent of the size of the machine.
2. Larger machines are particularly well suited for offshore wind power. The cost of foundations does not rise in proportion to the size of the machine.

Reasons for Choosing Smaller Turbines

1. The local electrical grid may be too weak to handle the electricity output from a large machine. This may be the case in remote parts of the electrical grid with a low population density and little electricity consumption in the area.
2. The cost of using large cranes, and building a road strong enough to carry the turbine components may make smaller machines more economic in some areas.
3. Aesthetic landscape considerations are dependent upon the perception of local people. Some people like to see larger machines and attach them with their own identity. Another set of people may not want to see any extraordinary large

H.-J. Wagner and J. Mathur, *Introduction to Wind Energy Systems*, Green Energy and Technology, https://doi.org/10.1007/978-3-319-68804-6_5

structures in the natural landscape. Speed of rotation also plays an important role in perception about landscape, since large machines have a much lower rotational speed, which means that one large machine really does not attract as much attention as many small, fast moving rotors.

In countries like Germany, where the availability of a strong grid to accept power is not a problem, the question of a small or large rotor area becomes insignificant. There is a clear cut tendency to prefer larger machines due to the economic advantage with a large rotor diameter. However, the selection between large and small machines becomes an important criterion in developing countries like India where the selection of larger machines may become inappropriate in the absence of availability of a strong grid. However, recently, even in developing countries like India, due to economic advantages, preference towards medium to larger capacity machines is seen over past few years. Government had launched a special program in the year 2016, inviting owners of smaller machines to change to larger machines, offering them some incentives. However, use of very large machines, typically of the order of 5 MW, in these countries is still not seen due to challenges related to transportation, installation, grid connection infrastructure.

5.2 Number of Blades

In modern wind turbines, use of an even number of rotor blades is avoided. The most important reason is the stability of the turbine. A rotor with an odd number of rotor blades (and at least three blades) can be considered to be similar to a disc when calculating the dynamic properties of the machine. A rotor with an even number of blades will give stability problems for a machine with a stiff structure. The reason is that at the very moment when wind is coming from the front, the uppermost blade bends backwards, because it gets the maximum power from the wind. If an even number of blades is used, at that time the lowermost blade passes into the wind shade in front of the tower. This combination makes the dynamic balancing of rotor more typical.

Most modern wind turbines, therefore, are three-bladed designs. The vast majority of the turbines sold in world markets have this design. Selection of the number of blades depends upon the wind profile.

Besides the above mentioned disadvantage, two-bladed wind turbine designs have the advantage of saving the cost of one rotor blade and its weight, of course. However, they tend to have difficulties in penetrating the market, partly because they require higher rotational speed to yield the same energy output. This is a disadvantage both in regard to noise and visual intrusion. Lately, several traditional manufacturers of two-bladed machines have switched to three-bladed designs.

One-bladed wind turbines also exist as prototypes, and indeed, they save the cost of another rotor blade. One-bladed wind turbines are not very widespread commercially, however, because the same problems that are mentioned with the

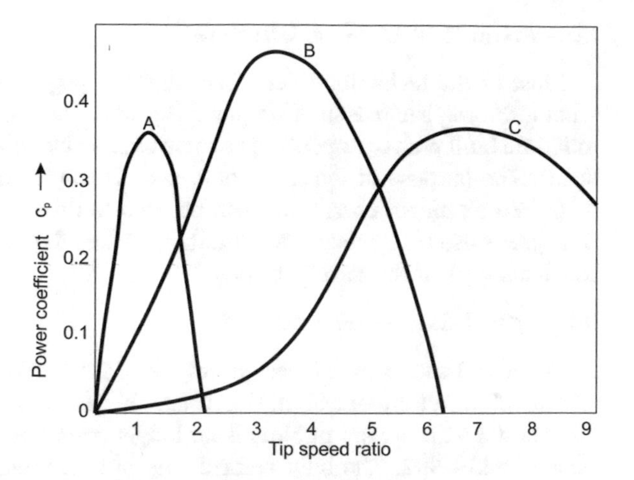

Fig. 5.1 Power coefficients and tip-speed ratio of different types of rotors (A: Savonius rotor, B: Three blade rotor, C: Two blade rotor)

two-bladed design apply to an even larger extent to one-bladed machines. In addition to the higher rotational speed, and the noise and visual intrusion problems, they require a counterweight to be placed on the other side of the hub from the rotor blade in order to balance the rotor. This obviously negates the savings on weight compared to a two-bladed design.

Two- and one-bladed machines require a more complex design with a hinged (teetering hub) rotor. The rotor has to be able to tilt in order to avoid too heavy shocks to the turbine when a rotor blade passes the tower. The rotor is, therefore, fitted onto a shaft which is perpendicular to the main shaft, and which rotates along with the main shaft. This arrangement may require additional shock absorbers to prevent the rotor blade from hitting the tower.

Due to mechanical problems, two blade and single blade designs have not become very popular. Additionally, it is understood that the three blade design is psychologically more acceptable to human perception as compared to the two or single blade design. Figure 5.1 shows the range of power coefficient and tip speed ratio for different types of rotors. It shows that as compared to the three blade rotors, the two blade rotors need a higher tip-speed ratio for operating at the same value of power coefficient. This means that for a given wind velocity, a two blade rotor has to rotate faster as compared to a three blade rotor.

5.3 Horizontal or Vertical Axis Turbine

Considering the present status of trade, the choice between horizontal axis and vertical axis wind turbines is almost theoretical and goes in favor of horizontal axis turbines. However research is going on to bring back vertical axis turbines into practice, which may be possible in future.

(a) **Horizontal Axis Wind Turbines**

Most of the technology described on these pages is related to horizontal axis wind turbines. The reason is simple: All grid-connected commercial wind turbines today are built with a propeller-type rotor on a horizontal axis (i.e. a horizontal main shaft). The purpose of the rotor, of course, is to convert the linear motion of the wind into rotational energy that can be used to drive a generator. The same basic principle is used in a modern water turbine, where the flow of water is parallel to the rotational axis of the turbine blades.

(b) **Vertical Axis Wind Turbines**

Figure 5.2 shows the biggest prototype of a 4.2 MW vertical axis wind turbine, with a 100 m rotor diameter at Cap Chat, Québec, Canada. The machine (which is the world's largest wind turbine) is no longer operational. It was operational in the period 1983–1992. The only vertical axis turbine which has ever been manufactured commercially at any volume is the Darrieus machine, named after the French engineer Georges Darrieus who patented the design in 1931. It was manufactured by the U.S. company FloWind which stopped its production in 1997. The Darrieus machine is characterized by its C-shaped rotor blades. It is normally built with two or three blades. Another type of vertical axis machine is the machine with H-rotor.

Fig. 5.2 A 4.2 MW vertical axis Darrieus wind turbine (*source* © www.reuk.co.uk)

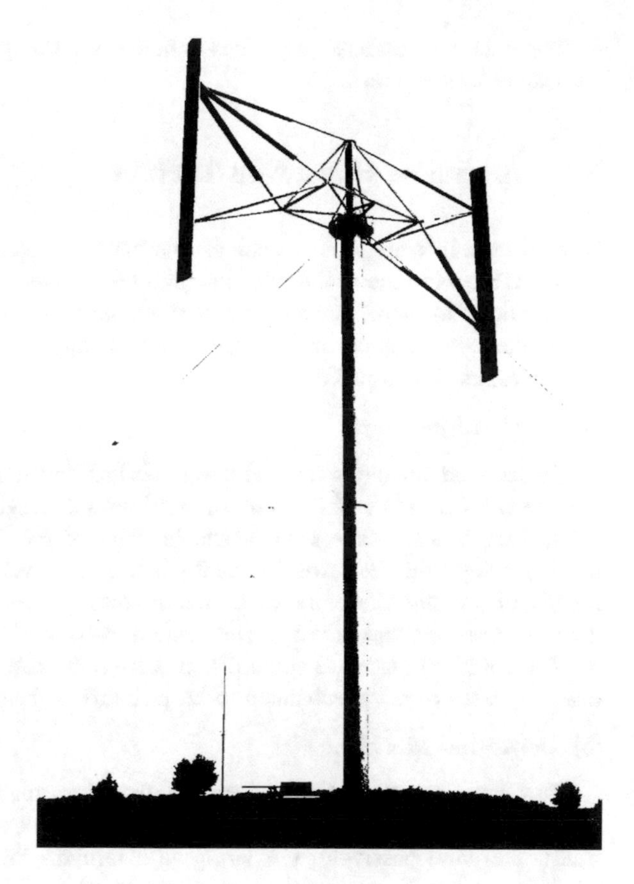

Fig. 5.3 H-rotor in Germany (*photo* © Große Böckmann)

The name is derived from the shape 'H' of its rotor (see Fig. 5.3). One more type of vertical axis machine consists of Savonius rotors which are primarily used for measuring weather conditions.

The basic theoretical advantages of a vertical axis machine are:

1. The generator, gearbox etc. can be placed on the ground, and may not need a tower for the machine.
2. A yaw mechanism is not needed to turn the rotor against the wind.

The basic disadvantages are:

1. Wind speeds are very low close to ground level, so the need of a tower is eliminated, but the wind speeds will be very low on the lower part of the rotor.
2. The overall efficiency of the vertical axis machines is less than in the case of a horizontal axis.
3. The machine is not self-starting, e.g. a Darrieus machine will need a "push" before it starts. This is only a minor inconvenience for a grid connected turbine, however, since the generator may be used as a motor drawing current from the grid to start the machine.

4. The machine may need guy wires to hold it up, but guy wires are impractical in heavily farmed areas.

5.4 Upwind or Downwind Turbine

Upwind machines are those machines that have the rotor facing the wind. In these machines the wind meets the rotor first and then leaves from the direction in which the nacelle is located. Downwind machines have the rotor placed on the leeward side of the tower; this means the nacelle comes first in the path of the wind and then the blades, as shown in Fig. 5.4.

(a) **Upwind Machines**

The basic advantage of upwind designs is that one avoids the wind shade behind the tower. By far the vast majority of wind turbines have this design. On the other hand, there is also some wind shade in front of the tower, i.e. the wind starts bending away from the tower before it reaches the tower itself, even if the tower is round and smooth. Therefore, each time the rotor passes the tower, the power from the wind turbine drops slightly. The basic drawback of upwind designs is that the rotor needs to be placed at some distance from the tower. In addition, an upwind machine needs a yaw mechanism to keep the rotor facing the wind.

(b) **Downwind Machines**

They have the theoretical advantage that they may be built without a yaw mechanism, if the rotor and nacelle have a suitable design that makes the nacelle follow the wind passively. For large wind turbines this is a somewhat doubtful advantage, since for optimal energy efficiency of wind energy converters, the yaw control must be applied very accurately.

Another advantage of the downwind design is that the rotor may be made more flexible. This is an advantage both in regard to weight and the structural dynamics

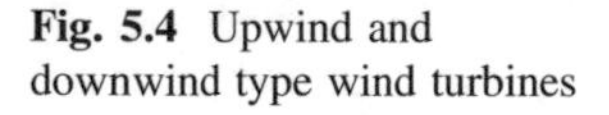
Fig. 5.4 Upwind and downwind type wind turbines

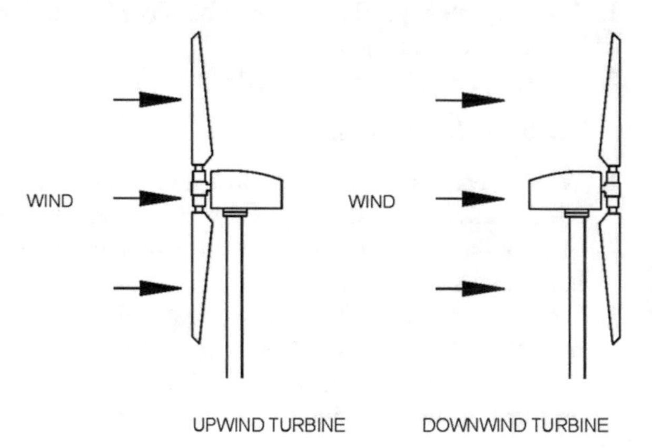

of the machine, i.e. the blades will bend at high wind speeds, thus taking part of the load off the tower.

The basic drawback is the fluctuation in the wind power due to the rotor passing through the wind shade of the tower. This may give more fatigue loads on the turbine than with an upwind design.

Out of these two options, upwind machines are more commonly used due to an increased energy output and hence their being more economical.

5.5 Load Considerations for Turbine Selection

Wind turbines are subject to fluctuating winds, and hence fluctuating forces. This is particularly the case if they are located in a very turbulent wind climate. Components which are subject to repeated bending, such as rotor blades, may eventually develop cracks which ultimately may make the component break. Metal fatigue is a well known problem in many technical goods. Due to combined reasons of fatigue and mass, metal is, therefore, generally not preferred as a material for rotor blades. When designing a wind turbine it is extremely important to calculate in advance how the different components will vibrate, both individually, and jointly. It is also important to calculate the forces involved in each bending or stretching of a component. This is the subject of structural dynamics, where mathematical computer models have been developed that analyse the behavior of an entire wind turbine. These models are used to design the machines safely.

As an example, a tall wind turbine tower will have a tendency to swing back and forth, say, every three seconds. The frequency with which the tower oscillates back and forth is also known as the eigenfrequency of the tower. The eigenfrequency depends on both the height of the tower, the thickness of its walls, the type of steel, and the weight of the nacelle and rotor. Now, each time a rotor blade passes the wind shade of the tower, the rotor will push slightly less against the tower.

If the rotor turns with a rotational speed such that a rotor blade passes the tower each time the tower is in one of its extreme positions, then the rotor blade may either dampen or amplify (reinforce) the oscillations of the tower. The rotor blades themselves are also flexible, and may have a tendency to vibrate, say, once per second. Consequently, it is very important to know the eigenfrequencies of each component in order to design a safe turbine that does not oscillate out of control.

5.6 Wind Turbines: With or Without Gearbox

(a) **Design with Gearbox**

The principle of a design of a wind turbine with gearbox, is shown in Figs. 5.5. and 5.6. The main aspect of this design is the split shaft system, where the main

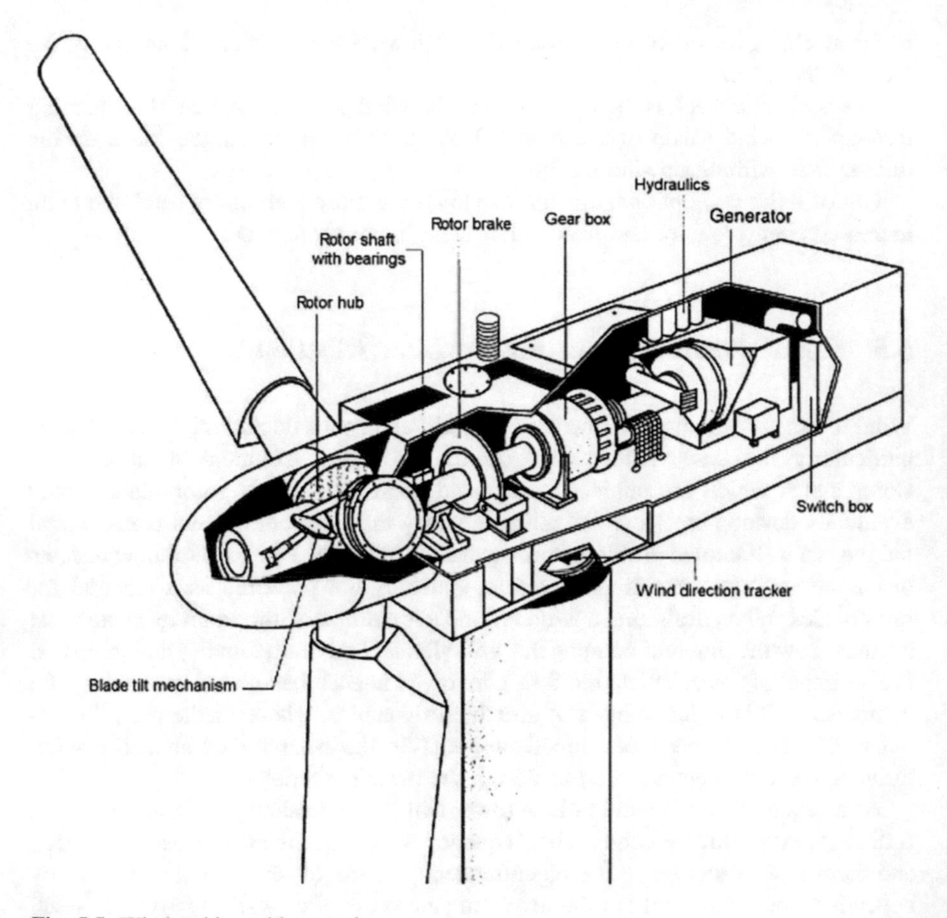

Fig. 5.5 Wind turbine with a gearbox

shaft turns slowly with the rotor blades and the torque is transmitted through a gearbox to the high-speed secondary shaft that drives the few-pole pair generator. The transmission of torque to the generator is shut off by means of a large disk brake on the main shaft. A mechanical system controls the pitch of the blades, so pitch control can also be used to stop the operation of the converter, e.g. in stormy conditions. The pitch mechanism is driven by a hydraulic system, with oil as the popular medium. For constructions without a main brake, each blade has its pitch angle controlled by a small electric motor.

The gearbox concept was in many cases accompanied by an insufficient life time because of failure of gearboxes. After many years of operational experiences and a lot of research and development activities it got solved.

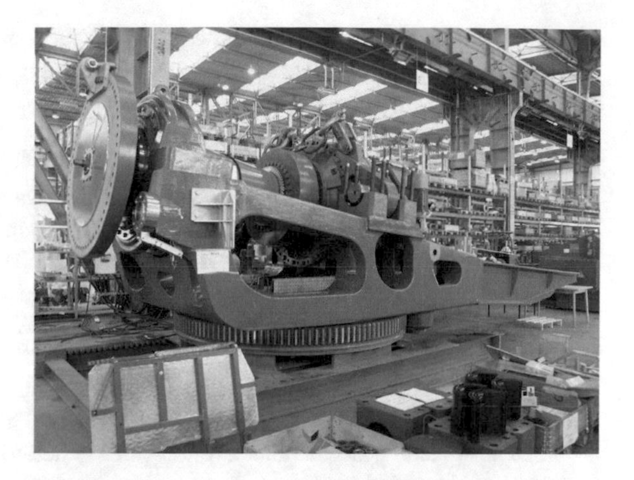

Fig. 5.6 Assembling of a wind turbine with gearbox by NORDEX company (*photo* © Nordex)

(b) **Design Without Gearbox**

Some companies, e.g. the German company Enercon, designed another converter type without gearbox. The scheme of such a converter is shown in Fig. 5.7, where the main design aspects can be clearly seen. This design has just one stationary shaft. The rotor blades and the generator are both mounted on this shaft. The generator is in the form of a large spoked wheel with e.g. forty-two pole pairs around the outer circumference and stators mounted on a stationary arm around the wheel. The wheel is fixed to the blade apparatus, so it rotates slowly with the blades. Therefore, there is no need for a gearbox, rotating shafts or a disk brake. This minimizing of rotating parts reduces maintenance and failure possibilities and

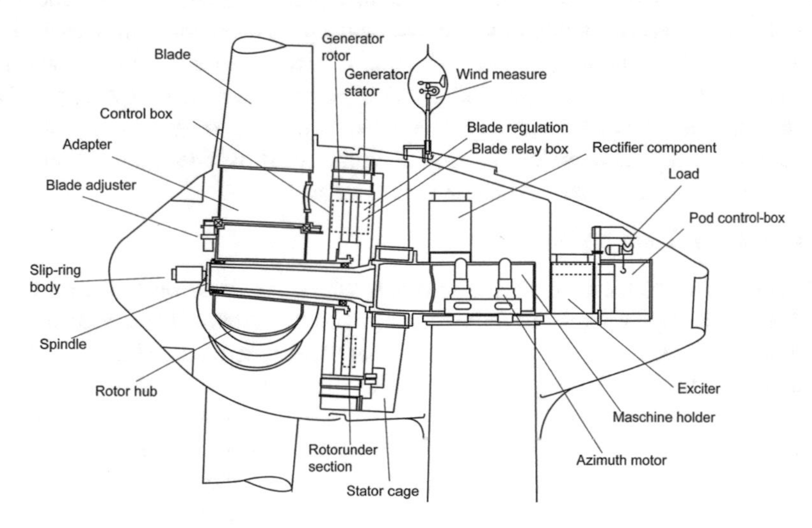

Fig. 5.7 Wind turbine without gearbox (design of ENERCON company)

simplifies the maintenance and production of the converter. The price for this advantages is a high nacelle mass caused by the high cupper content of multipole generator.

(c) **Design with Gearbox and Multipole Generator**

Both the designs explained above have one disadvantage each. The design with gearbox has the disadvantage of losses during transmission of power and high speed drive required for connecting the generator, while the gearless design has the disadvantage of increased weight of the nacelle due to an increased number of poles. The objective of this design is to overcome disadvantages of both previous designs. In this design, a new combination of approved techniques has been used. It was first operated in the M5000 design of the research and development company Multibrid (a registered name) about 10 years ago. This design is a combination between a special gearbox and a multi-pole generator. There is no high-speed transmission included in the gearbox, which is critical to failure. The selection of transmission achieves a high transmission allowing the use of a generator with up to for example 150 rpm, the generator has 10 poles (5 pole pairs). Additionally, in the prototypes, a permanent magnet-excited synchronous generator with water-cooling, high efficiency and wide speed range was used. Due to the compact design, the construction has a reduced weight.

The development has been tested successfully. Some producers in the market are offering with such kind of wind converters with success.

(d) **WinDrive Design**

In case of the WinDrive Design (registered name of the Voith Company) a hydrodynamic transmission with variable speed is used. The WinDrive, a torque converter, is placed between the main gearbox and the generator. This torque converter consists of three main components: pump wheel, turbine wheel and adjustable guide vanes. In this way variable input speeds and constant output speeds are possible. The generator converts the variable Input speeds constantly with 4 poles (2 pole pairs) in 1,500 rpm with 50 Hz line frequency. The advantage of this design is that the system is directly connected to the power supply. There are no failure prone power conversion electronics in the wind turbine. Installation of this type are e.g. in US, Canada, Argentina and China.

(e) **Flight Wind Energy Systems**

In contrast to conventional wind turbines, which are fixed on a tower or mast, flight wind energy systems are only connected to the ground via a rope or cable. Attempts are made to use the permanent higher wind speed at heights up to 450 m, resulting in a higher power generation of the wind power plants. A distinction is made here between a balloon type wind turbine and a type of steering kite. The balloon type wind turbine is filled with gases lighter than air, causing the system to float. Through the segments, the system rotates around the longitudinal axis and drives a generator. By the flight wind energy system, which is similar to a kite, the wind energy is tried to be used through a fixed wing or a fixed sail. In this case, the wing flies at maximum rope force in eighth direction across the wind. The rope is then released and drives a generator to produce electricity on the ground. When the highest point is reached, the wing slides quickly and with little force to the initial height, from where the wing returns back to the working phase. Currently, due to several problems, both systems are still in the test phase and only a few pilot plants exist.

5.7 Requirement of Grid, Synchronous or Asynchronous Generators

The main electrical grid has a constant frequency, e.g. 50 Hz or 60 Hz, and a constant phase angle. Therefore, a wind energy converter must produce electricity with the same constant values in order to integrate with electricity in the main grid. The input energy of a wind energy converter is proportional to the wind speed, but the wind speed is never constant. Each wind speed has a corresponding rotor rotation speed at which the maximum power is produced, as shown in Fig. 5.8. This maximum production of power occurs at different rates of rotation for different wind speeds. However, the rate of rotation must be kept constant in order to achieve the required constant output frequency, e.g. by 18.5 rpm in the figure. Solutions to this problem of maximizing the power output of converters by variable speeds and constant frequency are discussed in connection with the generator.

With respect to their working principle, there are two choices for generators: synchronous generators or asynchronous generators.

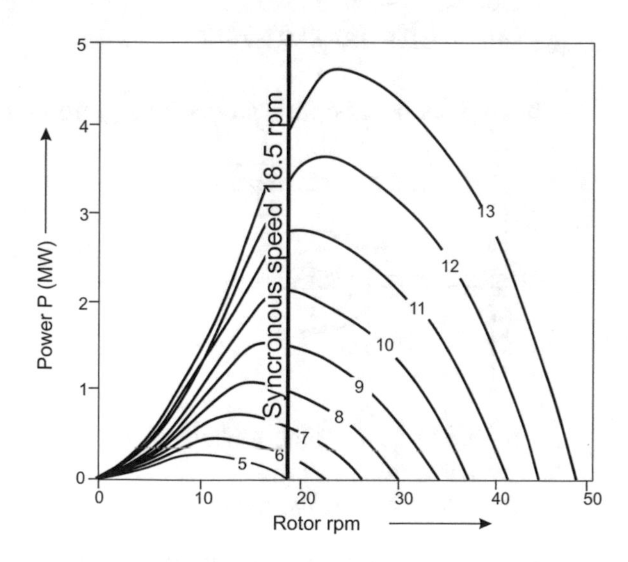

Fig. 5.8 Change in optimum rotation with wind speed for maximum power (synchronous speed 18.5 rpm)

(a) **Synchronous Generators**

All generators use a rotating magnetic field for the generation of electricity. The more force (torque) is applied and the rotational speed is kept constant, the more electricity is generated, but the synchronous generator will still run at the same speed dictated by the frequency of the electrical grid.

If the generator is disconnected from the main grid, however, it will have to be cranked at a constant rotational speed in order to produce alternating current with a constant frequency. The synchronous generator is generating power with constant frequency. Consequently, with this type of generator it is normally recommended to use an indirect grid connection for being able to operate it with different rotational speeds.

Figure 5.9 shows the connection of a wind turbine with synchronous generator to the grid. The synchronous generator operates with a variable speed of rotation,

a) Grid connection for synchronous generators with gear box

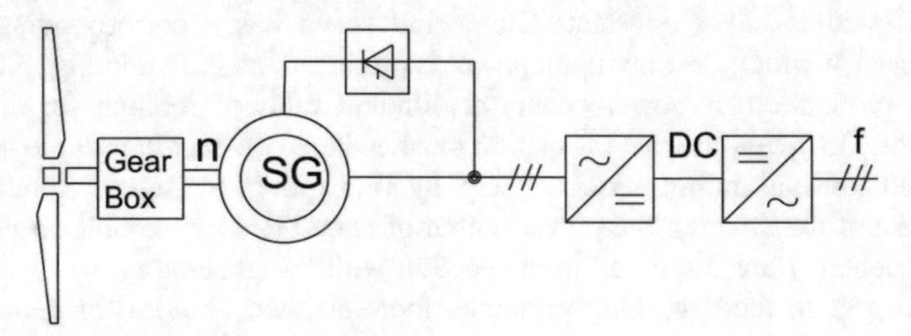

n= 0.5 to 1.2 (f/p) (controllable)
Inductive reactive load
Controllable reactive power

b) Grid connection for synchronous generators without gear box

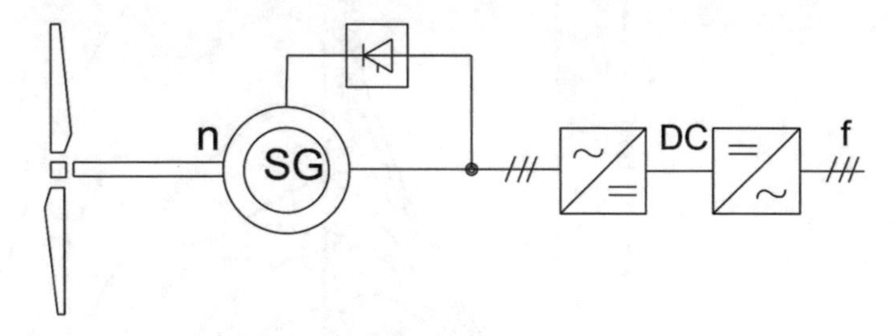

n= 0.5 to 1.2 (f/p) (controllable)
Inductive reactive load
Controllable reactive power

Fig. 5.9 Grid connection of synchronous generators (SG) with full inverter system

which means that it will produce electric power with variable frequencies. Due to a DC link, this will be transformed to AC power with the required frequency (usually the grid frequency). Use of DC has no link with the use of design having gearbox or without gearbox as shown in Fig. 5.9a, b for both cases.

b) **Asynchronous (Induction) Generators**

The other option for choosing a generator is a three phase asynchronous (cage wound) generator, also called an induction generator to generate alternating current. This type of generator is widely used in the wind turbine industry and in small hydropower units. The unique feature of this type of generator is that it was really originally designed as an electric motor. One reason for choosing this type of generator is that it is very reliable, and tends to be comparatively inexpensive with the basic design of a cage rotor. The generator also has some mechanical properties which are useful for wind turbines such as a generator slip, and a certain overload capability.

The speed of the asynchronous generator must be little higher than the synchronous speed. In practice, the difference between the rotational speed at peak power and at idle is very small, up to 8%. This difference in per cent of the synchronous speed is called the generator's slip (s). Thus, a 4-pole generator (2 pol-pairs (p)) will run idle at 1,500 rpm if it is attached to a grid with a 50 Hz current. If the generator is producing at its maximum power, let say with 1% slip, it will be running at 1,515 rpm. It is a very useful mechanical property that the generator will increase or decrease its speed slightly if the torque varies. This means that there will be less tear and wear on the gearbox due to lower peak torque. This is one of the most important reasons for using an asynchronous generator rather than a synchronous generator on a wind turbine which is directly connected to the electrical grid. The rotation of such direct grid connected asynchronous generators varies in the range of 100% of synchronous speed to 108% as shown in Fig. 5.10a.

In the section about the synchronous generator it has been discussed that it could run as a generator without connection to the public grid. An asynchronous generator is different, because it requires the stator to be magnetized from the grid before it works. An asynchronous generator can also run in a stand alone system, however, if it is provided with capacitors, which supply the necessary magnetization current. It also requires that there be some residual in the rotor iron, i.e. some leftover magnetism when the turbine is started. Otherwise a battery and power electronics or a small diesel generator is required to start the system. But it cannot run with constant frequency because the frequency is, as mentioned before, dependent on the rpm, which is dependent upon the generated electric power. In addition, an asynchronous generator can provide a limited amount of reactive power to the grid.

As discussed in the previous section, the motor (or generator) slip in an asynchronous (induction) machine is usually very small for reasons of efficiency, so the rotational speed will vary by up to 8% between idle and full load. The slip, however, is a function of the resistance in the rotor windings of the generator. So, one way of varying the slip is to vary the resistance in the rotor. In this way one may

Fig. 5.10 Grid connection of asynchronous generators (ASG)

a) Direct Grid Connection

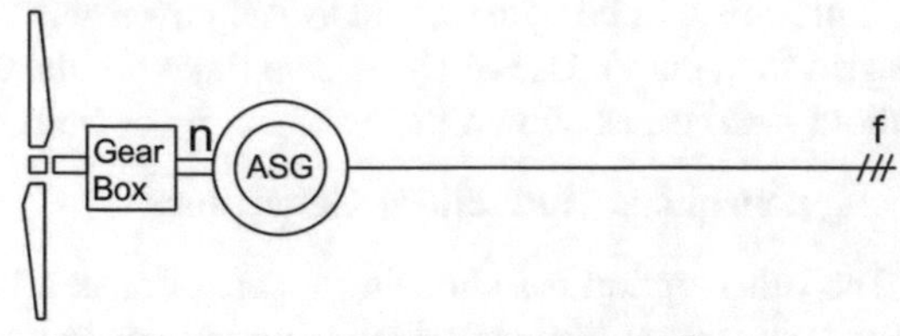

n= (1-s) f/p, s= 0 to 0.08 (Power Dependent)
Inductive reactive load

b) Dynamic Slip Control

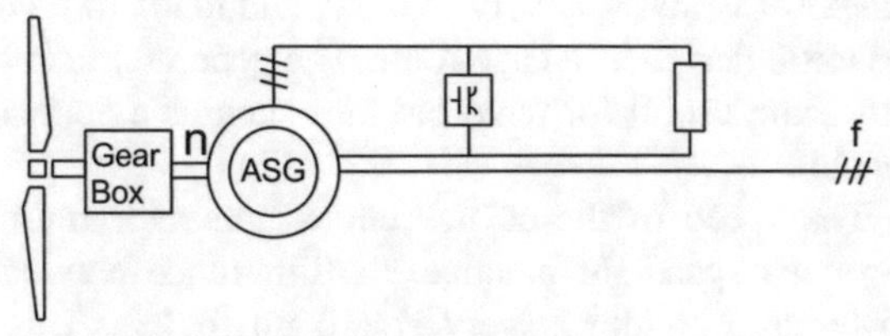

n= (1-s) f/p, s= 0 to 0.1 or 0.3 (Power Dependent, Dynamic)
Inductive reactive load

c) Double Excited Asynchronous Generator

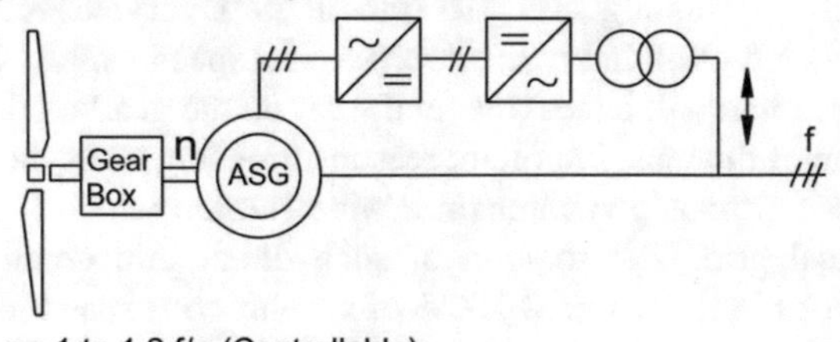

n= 1 to 1.2 f/p (Controllable)
Controllable Reactive Power

increase the generator slip to e.g. 10%. This can be done by having a wound rotor, i.e. a rotor with copper wire windings which are connected in a star, and connected with external variable resistors, plus an electronic control system to operate the resistors. The connection has usually been done with brushes and slip rings, which is a clear drawback over the elegantly simple technical design of a cage wound rotor machine. It also introduces parts which wear down in the generator. The rotation of such an asynchronous generator varies in the range of 100–110% of synchronous speed as shown in Fig. 5.10b.

Modern machines nowadays have double excited asynchronous generators as shown in Fig. 5.10c. In this case the generator is equipped with a winding in rotor and is excited with AC having variable frequency. This AC is produced by a DC link which is fed by the grid. This type of generators has an extended slip and is able to operate with rotational speed in the range of 100–120% of the synchronous

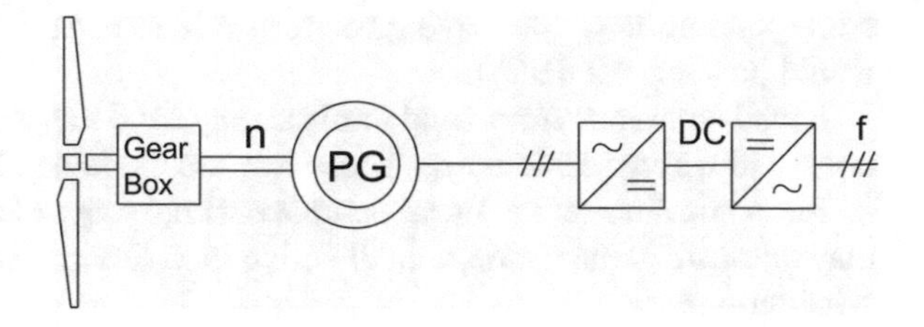

Fig. 5.11 Grid connection of a generator with permanent magnets (PG)

speed. A further advantage of such generators is that they can supply more reactive power to the grid as compared to the other two types discussed above.

(c) **Permanent Generators**

As mentioned earlier permanent magnet generators have the advantage of weight saving. Their development has been so encouraging that some companies have started equipping their new wind converters with permanent magnet generators. They do not need electricity for excitation (Fig. 5.11). An economical disadvantage could be the high costs of permanent magnets with high magnetic power. Some materials ideally suitable for applications, such as Neodynium, are scarcely available and are expensive too. The rotation of generators with permanent magnets can be adapted to the grid frequency in the same way as synchronous generators.

5.8 Issue of Noise and its Control

The loudness or strength of a noise or sound signal is measured with the unit 'decibel' abbreviated as 'dB'. One decibel is considered the smallest difference in sound level that the human ear can distinguish. Created in the early days of telephony as a way to measure cable and equipment performance and named after Alexander Graham Bell, decibels (dB) are a relative measurement derived from two signal levels: a reference input level and an observed output level. If a reference input level is used, the unit is often written as dB(A). A decibel is the logarithm of the ratio of the two levels. This means that if the dB level is doubling, the loudness level would be increasing by 10 times. A whisper is about 20 dB. A normal conversation is typically from 60 to 70 dB, and a noisy factory from 90 to 100 dB. Loud thunder is approximately 110 dB, and 120 dB borders on the threshold of pain.

The energy in sound waves (and thus the sound intensity) will drop with the square of the distance to the sound source, which is popularly known as the inverse square law. In other words, if one moves 200 m away from a wind turbine, the sound level will generally be one quarter of what it is 100 m away. To give an example: At one rotor diameter distance (43 m) from the base of a wind turbine emitting 100 dB(A) there will be generally a sound level of 55–60 dB(A) like a conversation. Four rotor diameters (170 m) away there will be 44 dB(A), corresponding to a quiet living room in a house. At 6 rotor diameters (260 m) it would be some 40 dB(A).

Sound emissions from wind turbines may have two different origins: Mechanical noise and aerodynamic noise. These two are explained below.

Mechanical noise, i.e. metal components moving or knocking against each other may originate in the gearbox, in the drive train (the shafts), and in the generator of a wind turbine.

When going by car, plane, or train, one may experience how resonance of different components, e.g. in the dashboard of a car or a window of a train may amplify noise. An important consideration, which enters into the turbine design process today, is the fact that the rotor blades may act as membranes that may retransmit noise vibrations from the nacelle and tower.

While designing, with the help of computer models it is ensured that the vibrations of different components do not interact to amplify noise. For example the chassis frame of the nacelle, on some of the wind turbines today, has some odd holes which are drilled into the chassis frame for no apparent reason. These holes were precisely made to ensure that the frame will not vibrate in step with the other components in the turbine.

When the wind hits different objects at a certain speed, it will generally start making a sound. If it hits the leaves of trees and bushes, or a water surface it will create a random mixture of high frequencies, often called white noise. The wind may also set surfaces in vibration, as sometimes happens with parts of a building, a car or even an engine-less glider plane. These surfaces in turn emit their own sound. If the wind hits a sharp edge, it may produce a pure tone.

In case of wind turbines, wind hits the rotor blades. Rotor blades make a slight swishing sound which may be heard close to a wind turbine at relatively low wind speeds. For their operation, rotor blades must split the wind on both sides of the blade so that energy can be transferred to the rotor. In the process they cause some emission of white noise. If the surfaces of the rotor blades are very smooth (which indeed they must be for aerodynamic reasons), the surfaces will emit a minor part of the noise. Most of the noise will originate from the trailing (back) edge of the blades. Careful design of trailing edges and very careful handling of rotor blades while they are mounted, have become routine practice in the industry for the purpose of controlling aerodynamic noise. Other things being equal, sound pressure will increase with the sixth power of the speed of the blade relative to the surrounding air. It can, therefore, be noticed that modern wind turbines with large rotor diameters have a very low rotational speed.

Since the tip of the blade moves substantially faster than the root of the blade, great care is taken about the design of the rotor tip. On close look at different rotor blades, it can be discovered that there has been substantial change in their geometry over time, as more and more research in the area has progressed. The research is also done for performance reasons, since most of the torque (rotational moment) of the rotor comes from the outer part of the blades. In addition, the airflows around the tip of rotor blades are extremely complex, compared to the airflow over the rest of the rotor blade.

One simple way of controlling noise is to reduce the tip-speed ratio by designing the rotor accordingly. An extension of this approach is sometimes used in Germany, by using different rotational speeds in day and night time due to different maximum permissible levels of noise. Many turbines can be seen with specially designed winglets at the tip of blades that help in reducing the noise, too, by controlling the turbulence at the tip.

Literatures

Heier S (2006) Grid integration of wind energy conversion systems. Wiley, London. ISBN 978-0470868997

Patel Mukund R (2006) Wind and solar power systems—design, analysis and operation. CRC Press, Boca Raton, FL. ISBN 0849315700

Chapter 6
Operation and Control of Wind Energy Converters

After having discussed the appropriate wind energy system design, it is equally important to discuss its optimal operation and control of performance. The present chapter covers such aspects.

6.1 Power Curve and Capacity Factor

6.1.1 Power Curve

The power curve of a wind turbine is a graph that indicates how large the electrical power output will be for the turbine at different wind speeds. Figure 6.1 shows the shape of a theoretical power curve of a wind turbine.

Below the main features of any power curve are described:

(a) **The Cut In Wind Speed (v_c)**

Usually, wind turbines are designed to start running at wind speeds somewhere around 3–5 m/s. This is called the cut in wind speed. Below this speed of wind, the energy in wind is not sufficient to overcome the inertia of the rotor; hence, the machine does not produce any power below this speed of wind.

(b) **The Cut Out Wind Speed (v_f)**

The wind turbine will be programmed to stop at high wind speeds above, say, 25 m per second, in order to avoid damaging the turbine or its surroundings. The stop wind speed is called the cut out wind speed.

(c) **Rated wind speed (v_r)**

The "rated wind speed" is the wind speed at which the "rated power" is achieved. This value for megawatt size turbines is about 12–15 m/s, and it

H.-J. Wagner and J. Mathur, *Introduction to Wind Energy Systems*, Green Energy and Technology, https://doi.org/10.1007/978-3-319-68804-6_6

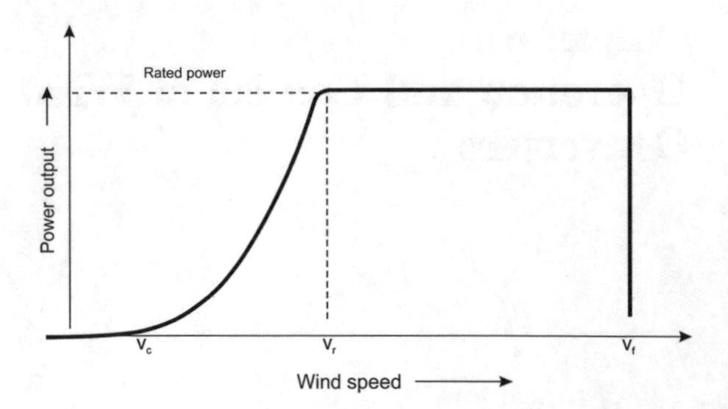

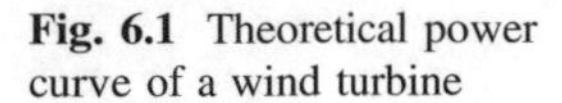

Fig. 6.1 Theoretical power curve of a wind turbine

corresponds to the point at which the conversion efficiency is near its maximum. The power output above the rated wind speed is mechanically or electrically maintained at a constant level, because the high output would destroy the equipment.

(d) **Survival speed**

Besides the above three important speeds of any power curve, there is one more speed which is specific together with the power curve: survival speed. It is that speed at which the wind converter would not be able to sustain for its survival. The survival speed being more than cut-off speed, is not shown in the power curve. Its value is around 50–60 m/s. This value becomes one very important factor when selecting a wind turbine. It should be ensured that the maximum ever wind velocity in that location is lower than the survival speed of the machine.

It can be noted from the power curve that at lower wind speeds, the power output drops off sharply. This can be explained by the cubic power law, which states that the power available in the wind increases eight times for every doubling of wind speed and decreases eight times for every halving of the wind speed, as discussed in earlier chapters.

Using the power curve, it is possible to determine roughly how much power will be produced at the average or mean wind speed prevalent at a site. However, it is recommended to use the Weibull distribution for estimating the power output in connection with the power curve.

Power curves are calculated and found by field measurements, where an anemometer is placed on a mast reasonably close to the wind turbine (not on the turbine itself or too close to it, since the turbine rotor may create turbulence, and make wind speed measurement unreliable). If the wind speed is not fluctuating too rapidly, then one may use the wind speed measurements from the anemometer and read the electrical power output from the wind turbine and plot the two values together in a graph. In practice the wind speed always fluctuates, and one cannot measure exactly the column of wind that passes through the rotor of the turbine. It is not a workable solution just to place an anemometer in front of the turbine, since the turbine will also cast a "wind shadow" and brake the wind in front of itself.

In practice, therefore, one has to take an average of the different measurements for each wind speed, and plot the graph through these averages. Furthermore, it is difficult to make exact measurements of the wind speed itself. If one has a 3% error in wind speed measurement, then the energy in the wind may be 9% higher or lower (remember that the energy content varies with the third power of the wind speed). Consequently, there may be some errors in certified power curves. Nevertheless, a power curve is a very useful guide for estimating the output from a wind turbine.

6.1.2 Capacity Factor

Capacity factor is one element in measuring the productivity of a wind turbine or any other power production facility and to compare different locations with each other. It compares the plant's actual production over a given period of time, e.g. one year, with the amount of power the plant would have produced if it had run at full capacity for the same amount of time.

A conventional utility power plant, unless used as a peak load power plant, will normally run during most of the time unless it is idle only due to equipment problems or for maintenance or due to a reduced demand of energy. A capacity factor of 40–80% is typical for conventional plants. A wind plant is "fueled" by the wind, which blows steadily at times and not at all at other times. Although modern utility-scale wind turbines typically operate 65–90% of the time, they often run at less than full capacity. Therefore, a capacity factor of 20–40% is common over one year, although they may achieve higher capacity factors during windy weeks or months and lower capacity factors in windless duration.

In Germany, another term, i.e. 'load duration' is used to indicate the capacity factor. The term is a multiple of the total number of hours in one year (8,760 h) and the capacity factor. Since the capacity factor is much lower than one, the load duration also comes out to be much less than 8,760. For example, if the capacity factor is 20%, the load duration would be 8,760 * 0.2, which is 1,752 h.

The significance of load duration is that it expresses that number of hours for which the wind turbine can be considered to be virtually operating at its rated capacity in one year.

Here it is important to distinguish between plant availability factor and plant capacity factor. The first term refers to a fraction of one complete year for which the plant is available for use, irrespective of the availability of wind. For example, if a plant is under maintenance for 200 h in one year, out of a total of 8760 h of one year, the availability factor would then be: (8,760–200)/8,760, i.e. 0.9771 or 97.71%.

To understand the capacity factor, let us consider a case of one 2 MW wind turbine:

In one year, a 2 MW wind turbine (8,760 h) can theoretically produce energy: 8,760 h * 2 MW = 17,520 MWh.

If it actually produced 4,000 MWh only due to fluctuations in wind availability, the ratio of these two would be the capacity factor calculated below:

Capacity factor = 4,000/17,520 = 0.2283 or 22.83%.

It is important to note that while the capacity factor is almost entirely a matter of reliability for a fueled power plant, this is not the case with a wind plant. In case of a wind plant, it is a matter of economical turbine design. With a very large rotor and a very small generator, a wind turbine would run at full capacity whenever the wind blew and would have a 60–80% capacity factor, but it would produce very little electricity. The most electricity per unit of investment is gained by using a larger generator and accepting the fact that the capacity factor will be lower as a result. Wind turbines are fundamentally different from fueled power plants in this respect due to a limited and fluctuating availability of wind.

If a wind turbine's capacity factor is 33%, this does not mean it is only running one-third of the time. Rather, a wind turbine at a typical location would normally run for about 65–90% of the time. However, much of the time it will be generating at less than full capacity, making its capacity factor lower.

6.2 Power Control of Wind Turbines

Wind turbines are designed and operated to produce electrical energy as cheaply as possible. Wind turbines are, therefore, generally designed such that they yield a rated power output at wind speeds around 12–15 m per second. As discussed in previous sections, the exact value of such a wind velocity varies with manufacturer and size due to the difference in power curve. Its does not pay to design turbines that maximize their output at stronger winds, as such strong winds are rare. In case of stronger winds it is necessary to waste part of the excess energy of the wind in order to avoid damage to the wind turbine.

Figure 6.2 shows the electricity power curve of a 2.5 MW machine together with two more important curves: input power of wind together with the decline power coefficient (c_p) for wind velocities exceeding the rated velocity. The product of these two represents the net efficiency of the wind turbine. The deviation between the input power and the power curve is due to net efficiency '$\eta \cdot c_p$'. It can be very well observed that the curve for net efficiency reduces after reaching its maximum slightly before the rated wind speed. It is due to the fact that the pitch control starts wasting energy of wind after this point maintaining power under limits.

All wind turbines are designed with some sort of power control. There are two different ways of doing this safely on modern wind turbines.

6.2.1 *Pitch Control*

On a pitch controlled wind turbine the turbine's electronic controller checks the power output of the turbine several times per second. When the power output becomes too high, it sends an order to the blade pitch mechanism which

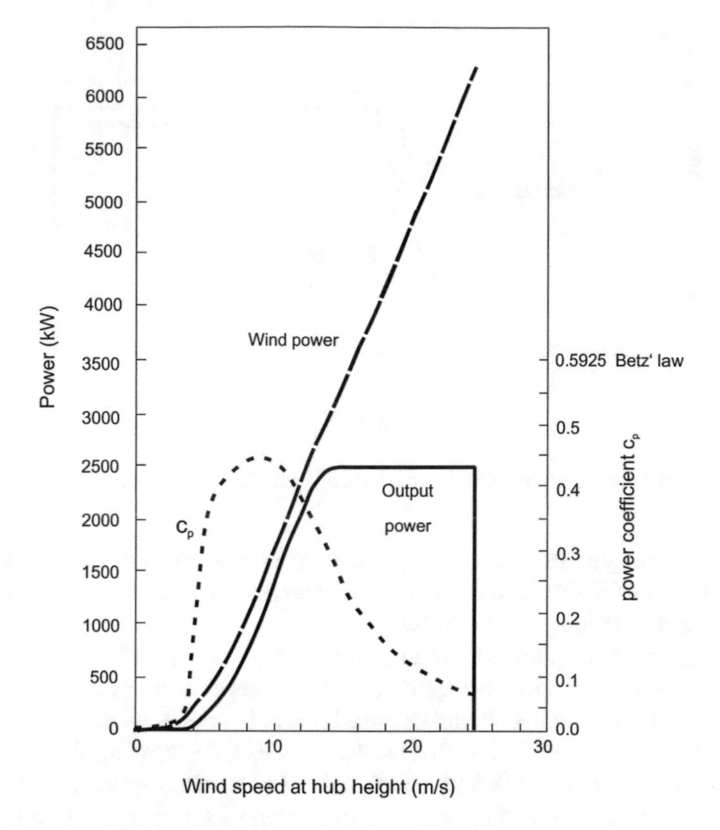

Fig. 6.2 Capacity utilization and power curve of a wind turbine

immediately pitches (turns) the rotor blades slightly out of the wind. Conversely, the blades are turned back into the wind whenever the wind drops again. The rotor blades thus have to be able to turn around their longitudinal axis (to pitch) as shown in Fig. 6.3.

During normal operation the blades will pitch a fraction of a degree at a time—and the rotor will be turning at the same time. The pitch mechanism can be operated

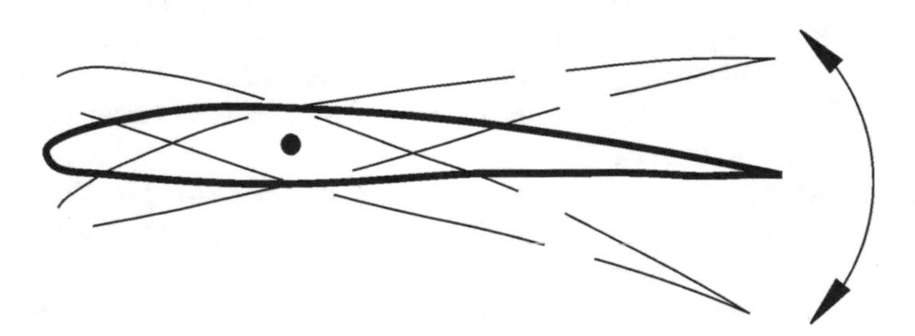

Fig. 6.3 Variable pitch blades

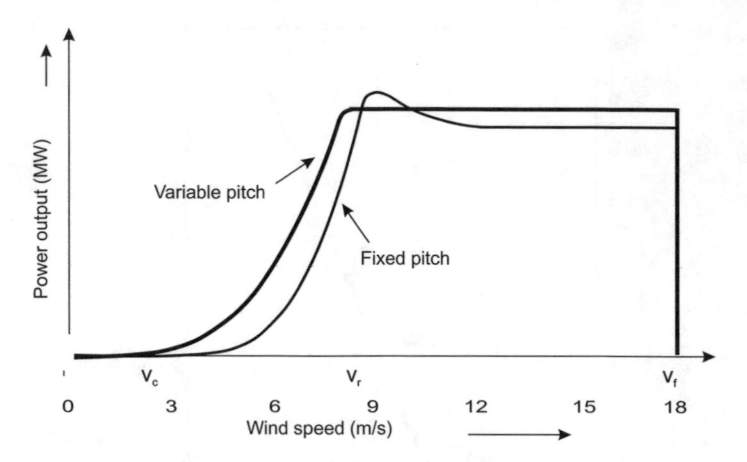

Fig. 6.4 Comparison of wind turbines with fixed and variable pitch control

using hydraulic systems. But in most cases, individual electric drives are used to actuate control of blades, and the same mechanism is also used for applying brakes to the rotor by turning just one or two blades.

The importance of pitch control is clearly visible in Fig. 6.4. In the fixed pitch machine, power output dips very quickly at a wind speed higher than the rated wind speed (v_r) and is oscillating around the rated power by higher wind speeds. This can be understood in connection with Fig. 3.5 in Chap. 3, showing the relationship between power coefficient and tip speed ratio. As the wind speed increases, for a fixed rate of revolution of the rotor, the tip speed ratio decreases. As a result, the power coefficient decreases very sharply, which reduces the power output from the wind mill sharply as well. Whereas, in pitch controlled machines, due to a change of the blade angle, the curve itself changes which avoids a sudden dip in the power coefficient and output power.

6.2.1.1 Running a Pitch Controlled Turbine at Variable Speed

There are a number of advantages of being able to run a wind turbine at variable speed. One good reason for wanting to be able to run a turbine partially at variable speed is the fact that pitch-control (controlling the torque in order not to overload the gearbox and generator by pitching the wind turbine blades) is a mechanical process. This means that the reaction time for the pitch mechanism becomes an important factor in turbine design.

In a variable slip generator, however, one may start increasing its slip once it is close to the rated power of the turbine. The control strategy is to run the generator at half of its maximum slip when the turbine is operating near the rated power. When a wind gust occurs, the control mechanism signals to increase the generator slip to allow the rotor to run a bit faster, while the pitch mechanism begins to cope with the situation by pitching the blades more out of the wind. Once the pitch mechanism has done its work, the slip is decreased again. In case the wind suddenly drops, the process is applied in reverse.

6.2.2 Stall Control

An aircraft wing will stall, if the shape of the wing tapers off too quickly as the air moves along its general direction of motion. The turbulence is created on the back side of the wing in relation to the air current. Stall can be provoked if the surface of the aircraft wing—or the wind turbine rotor blade—is not completely even and smooth. A dent in the wing or rotor blade, or a piece like a self-adhesive tape can be enough to start the turbulence, even if the angle of the wing in relation to the general direction of airflow is small. Aircraft designers obviously try to avoid stall at all costs, since an aero plane without the lift from its wings will start falling down. Similarly, stall condition arriving in a wind energy converter would result in a no output condition.

In order to avoid stall condition, the rotor blades for large wind turbines are always twisted. Seen from the rotor blade, the wind will be coming from a much steeper angle (more from the general wind direction in the landscape), as one moves towards the root of the blade, and the centre of the rotor. As discussed above for stall, a rotor blade will stop giving lift, if the blade is hit at an angle of attack which is too steep. This is an additional reason for twisting blades to achieve an optimal angle of attack throughout the length of the blade. However, in the case of stall controlled wind turbines in particular, it is important that the blade is built such that it will stall gradually from the blade root and outwards at high wind speeds.

6.2.2.1 Passive Stall Control

Passive stall controlled wind turbines have the rotor blades bolted onto the hub at a fixed angle. The geometry of the rotor blade profile, however, has been aerodynamically designed to ensure that the moment the wind speed becomes too high, it creates turbulence on the side of the rotor blade which is not facing the wind. This stall prevents the lifting force of the rotor blade from acting on the rotor.

The basic advantage of stall control over pitch control is that one avoids moving parts in the rotor itself, and a complex control system. On the other hand, stall control represents a very complex aerodynamic design problem, and related design challenges in the structural dynamics of the whole wind turbine, e.g. to avoid stall-induced vibrations. A normal passive stall controlled wind turbine will usually have a drop in the electrical power output for higher wind speeds, as the rotor blades go into deeper stall. Therefore, large wind machines are not using stall control for controlling output power.

6.2.2.2 Active Stall Control

An increasing number of larger wind turbines (1 MW and more) are being developed with an active stall power control mechanism. Technically the active stall machines resemble pitch controlled machines, since they have pitchable blades. In order to get a reasonably large torque (turning force) at low wind speeds, the machines will usually be programmed to pitch their blades much like a pitch controlled machine at low wind speeds. (Often they use only a few fixed steps depending upon the wind speed.) When the machine reaches its rated power, however, an important difference from the pitch controlled machines can be noticed. The difference is that if the generator is about to be overloaded, the machine will pitch its blades in the opposite direction from what a pitch controlled machine does. In other words, it will increase the angle of attack of the rotor blades in order to make the blades go into a deeper stall, thus wasting the excess energy in the wind.

One of the advantages of active stall is that one can control the power output more accurately than with passive stall, so as to avoid overshooting the rated power of the machine at the beginning of a gust of wind. Another advantage is that the machine can be run almost exactly at rated power at all high wind speeds.

6.2.3 Yaw Control

Some older wind turbines use flaps to control the power of the rotor, just like aircrafts use flaps to alter the geometry of the wings to provide extra lift at the time of takeoff. Another theoretical possibility is to yaw the rotor partly out of the wind to decrease power. This technique of yaw control is in practice used only for tiny wind turbines, as it subjects the rotor to cyclically varying stress which may ultimately damage the entire structure. The wind turbine yaw mechanism is used to turn the wind turbine rotor against the wind.

The wind turbine is said to have a yaw error, if the rotor is not perpendicular to the wind. A yaw error implies that a lower share of the energy in the wind will be running through the rotor area. The share will drop to the cosine of the yaw error, due to the change in the projected area of the rotor facing the wind.

If this were the only thing that happened, then yaw control would be an excellent way of controlling the power input to the wind turbine rotor. The main problem with yaw control is that the part of the rotor which is closest to the source direction of the wind, however, will be subject to a larger force (bending torque) than the rest of the rotor. On the one hand, this means that the rotor will have a tendency to yaw against the wind automatically, regardless of whether we are dealing with an upwind or a downwind turbine. On the other hand, it means that the blades will be bending back and forth in a flapwise direction for each turn of the rotor. Wind turbines which are running with a yaw error are therefore subject to larger fatigue loads than wind turbines which are yawed in a perpendicular direction against the wind.

6.3 Connection to the Grid

The connection to grid depends upon the purpose for which a wind energy system is used. Although most large wind energy systems are installed to feed electricity to the grid, however, other configurations are also found, particularly in small and medium size turbines. Their requirements for grid connection are described in this section.

6.3.1 Applications of Wind Energy Converters

From the point of view of utilization of generated electricity, wind energy systems can be used in three ways as explained below:

(a) **Grid connected system**

The grid connected system has a connection with an electricity transmission and distribution system called grid-connected systems. A grid-connected wind turbine can be used for reducing the consumption of fossil energy carriers. If the turbine cannot deliver the amount of energy needed, the public utility makes up the difference.

(b) **Stand-alone system**

As their name suggests, stand-alone systems are not connected to the utility grid. Stand-alone wind energy systems can be appropriate for homes, farms, or even entire communities (a co-housing project, for example) that are far from the nearest utility lines. However, if grid is available in a nearby area, it is always advantageous to be connected to the grid. Therefore, stand alone systems are used only in areas where grid is not available at all.

(c) **Hybrid systems**

A hybrid system using wind energy is one in which operation of a wind turbine is combined with any other source of power. Such a combination can be considered as another type of stand alone system. This source may be e.g. a photovoltaic system, or a diesel generator set. To specify the nature of these systems, sometimes the term 'stand-alone hybrid system' is also used to differentiate between the hybrid systems in which wind turbine is operating in hybrid mode with the grid supply.

At many sites, wind speeds are low in the summer when the sun shines brightest and longest. The wind is strong in the winter when there is less sunlight available. Since the peak operating times for wind and photovoltaic systems occur at different times of the day and year, hybrid systems are more likely to produce power when required.

For the times when neither the wind generator nor the photovoltaic modules are producing electricity (for example, at night when the wind is not blowing), most stand-alone systems provide power through batteries and/or an engine-generator powered by fossil fuels like diesel.

If the batteries run low, the engine-generator can be run at full power until the batteries are charged. Adding a fossil-fuel-powered generator makes the system more complex, but modern electronic controllers can operate these complex systems automatically. Adding an engine-generator can also reduce the number of photovoltaic modules and batteries in the system. It should be ensured that the storage capability is large enough to supply electrical needs during non-charging periods. Battery banks are typically sized for one to three days of windless operation.

6.3.2 Voltage Requirement

A single small converter can be directly connected into the grid network at e.g. 0.4 kV level. Once the wind energy converter is integrated into the grid network, there must be very limited voltage change, voltage oscillation or flicker experienced in the homes on that network branch. The loss of voltage due to resistance in the cabling can be avoided by increasing the diameter of the cables. It is often required that a new network branch is constructed and linked to the transformer in order to reduce the voltage disturbances. This increases the installation costs of the converter.

A single Mega-Watt size converter cannot be connected to the grid at the 0.4 kV stage, but has to be connected at 10–30 kV, which is the usual level of the city electricity share distribution. In remote areas, where a 30 kV connection is not established, the connection must be created and financed. Wind parks with a lot of Mega-Watt converters must be connected into the electrical grid at a level of about 100 kV and higher.

As mentioned earlier, the maximum power output is obtained only in a few hours during the year. Figure 6.5 shows a typical load distribution, measured within the German 250 MW program. With larger wind energy installations, this uneven distribution leads to the need of higher regulation capacities by conventional power systems in the future.

It would be worth mentioning here that due to fluctuations in availability of wind, these systems alone cannot fully meet the energy demand. On a large scale, additional conventional power plants that are flexible in their level of operation are needed. Whenever wind is available, the level of operation of the conventional

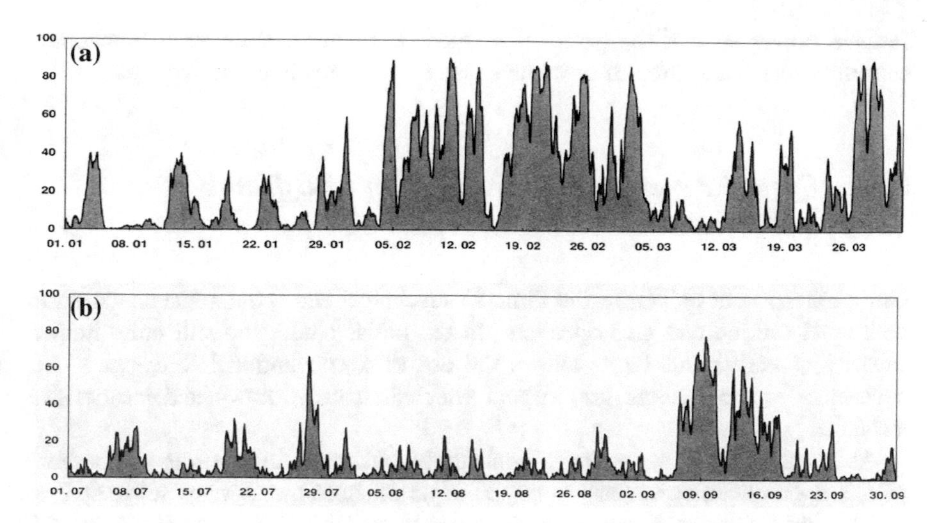

Fig. 6.5 Load distribution of a wind park with a total capacity of 28 MW in Germany. (**a**) July–September 1997. (**b**) January–March 1997 (*source* ISET, see literature)

power plant is reduced, and when wind is not available, it is increased to meet the demand.

From electrical point of view wind converter must be able to participate in the regulation and control of grid, especially when the share of wind energy in the grid electricity is high. Therefore bigger wind energy converters must be able to deliver different amounts of reactive power and they must be able to reduce the active power when the grid does not need so much of it. Due to a considerable number of wind turbine installations, for example wind turbines in Germany have to be adapted and controlled remotely to maintain the grid stability and the grid frequency of 50 Hz by law since 2009. Wind turbines of MW scale must adapt to changing frequency by changing their operation by about 10% reduction in power generation or load supplied per minute.

The grid voltage is different depending on the location and the local consumption. Although the voltage can be regulated by a feed-in or through the consumption of reactive power, today's wind turbine generators are decoupled from the power grid via a DC link. Due to presence of the DC link, it is also possible to control the phase shift between voltage and current in the three-phase current supplied. These systems no longer load the network with reactive power, but can be used by an excitation (inductive) or over-excitation (capacitive) for reactive power compensation.

In the event of a short circuit in the network, a regional potential gradient as well as an increased active power requirement arises. The voltage near the short-circuit point is equal to zero and at a great distance therefrom almost equal to grid voltage. A wind turbine close to the short circuit must be able to supply the highest possible reactive power so that the potential gradient is reduced. With such functioning capability, the wind turbines nowadays also serve to maintain the voltage.

6.3.3 Special Aspects of the Connection of Offshore Wind Parks

Grid connection of offshore wind farms is a technical and economical challenge to both wind turbine and grid operators. In the initial phase, the still quite limited capacity of early pilot farms allows the use of a conventional three-phase AC connection to the onshore grid system which is a well known and inexpensive technology.

An internal grid is necessary to connect the offshore wind farm to the onshore grid. The produced power has to be fed to an offshore transformer substation, to which wind turbines are connected via undersea cables by a voltage of e.g. 30 kV. After stepping-up to the transmission line voltage, the power is conveyed to the shore. The energy delivered to the coast via the undersea cable is usually fed into the power grid at the high voltage level. For transmission distances from offshore wind parks to the coast of approximately 50 km and more, High Voltage DC transmission is an alternative to AC transmission in offshore wind parks. This type of transmission results in less losses, since with DC no reactive power has to be transmitted. For this purpose, the AC of the wind turbine is converted into DC in a converter platform in the sea. Since the world's most current networks are AC networks, there is a converter at the end of each High Voltage DC transmission, which transfers the incoming DC into AC. As a result of the associated costs and losses in the inverter, the economic aspects of the electrical connection with the coast have to be compared between an AC and DC transmission.

Better suitability of submarine High Voltage DC transmission lines for offshore wind turbines is still a debatable issue. High Voltage DC lines offer the advantage of reduced transmission losses but they have two major problems, one is, of course, the cost, and the other is that their life with present materials and technologies in use, is relatively low as compared to high voltage AC transmission lines. The issue is much more important as compared to on-shore turbines due to greater lengths of transmission, which are of the order of 5–10 km.

Literatures

Im Auftrag des Bundesministeriums für Bildung, Wissenschaft, Forschung und Technologie, Kassel: Institut für Solare Energieversorgungstechniken (ISET) (in German language)

Scientific Measure and Evaluation Program for the 250 MW-Wind Test Program, Results of 1997 (1998). Wissenschaftliches Mess- und Evaluierungsprogramm (WMEP) zum Breitentest 250 MW Wind, Jahresauswertung

Chapter 7
Economics and Policy Issues

This chapter presents various economic and financial considerations that are important for selecting wind turbines, and before taking a decision in favor of installing a wind turbine at all.

7.1 Cost of Wind Turbines

7.1.1 Initial Cost of Wind Turbine

The average price for large, modern onshore wind farms in Germany is between 1,520 and 1,710 Euro per kilowatt electrical power installed, in India it is slightly less than 1,000 Euro (72,000 Indian Rupees) per kilowatt, mainly due to cheaper labor costs. These costs are just indicative figures and would differ with capacities and make. For single turbines or small clusters of turbines the costs will usually be somewhat higher as compared to large wind farms, due to bulk purchase price breaks. These costs also increase with an increase in tower height.

Installation costs include foundations, road construction (necessary to move the turbine and the sections of the tower to the building site), a transformer (necessary to convert the low voltage (e.g. 690 V) current from the turbine to 10–30 kV) current for the local electrical grid, telephone connection for remote control and surveillance of the turbine, and cabling costs, i.e. the cable from the turbine to the local 10–30 kV power line. In addition, transport to the construction site, the wind turbine itself, assembly, planning and approvals as well as environmental measures are added.

Table 7.1 shows the composition of an investment plan in €/kW as an example for the hub heights below 120 m and over 120 m and the capacity power of about 2–3 MW.

H.-J. Wagner and J. Mathur, *Introduction to Wind Energy Systems*, Green Energy and Technology, https://doi.org/10.1007/978-3-319-68804-6_7

Table 7.1 Investment plan for hub heights <120 m and >120 m in €/kW

Hub height	<120 m	>120 m
Wind turbine, transport to the building site, assembly	1,150	1,340
Foundation	70	
Grid connection	70	
Side development	40	
Planning, environmental measures, approval, other	190	
Total	1,520	1,710

The investment costs are different in the area of wind turbine, transport to the construction site and assembly. These depend on the hub height and rise with increasing hub height. The remaining investment points are approximately the same for different hub heights.

Obviously, the costs of roads and foundations depend on soil conditions, i.e. how cheap and easy it is to build a road capable of carrying heavy trucks. Another variable factor is the distance to the nearest ordinary road, the cost of getting a mobile crane to the site, and the distance to a power line capable of handling the maximum energy output from the turbine. Transportation costs for the turbine may enter the calculation, if the site is very remote. Of course this figure also depends upon the weight of the turbine, distance to be covered and, last but not least, current market prices.

It is obviously cheaper to connect many turbines at the same location, than just one. On the other hand, there are limits to the amount of electrical energy the local electrical grid can handle. Especially in developing countries like India, if the local grid is too weak to handle the output from the turbine, a need for grid reinforcement may arise, i.e. by extending the high voltage electrical grid. Agency that bears the cost of grid reinforcement: the power company or the turbine owner; varies from country to country.

7.1.2 Operation and Maintenance Costs for Wind Turbines

Modern wind turbines are designed to work for a design lifetime of 20 years and more.

Operation Maintenance Costs

Experience shows that maintenance costs are generally very low when the turbines are brand new, but they increase somewhat as the turbines age. The newer generations of turbines have relatively lower repair and maintenance costs that the older generations. Older wind turbines (upto 500 kW) require annual maintenance costs at an average of around 3% of the original turbine investment. Newer turbines are on average substantially larger, which would tend to lower maintenance costs per kW installed power, since it is not needed to service a large, modern machine more often than a small one. For newer machines the estimates range around 1.5–2% per year of the original turbine investment. In some cases, however, e.g. if the product is not standardized and/or the site is not investigated thoroughly, even these costs may be relatively high as compared to industry average.

The operational costs split up in maintenance and repairs, lease payments for land, operational management (as well technical as commercial), reserves for unforeseen events and for insurance. Table 7.2 shows the percentage of the operational costs of 5.0 €ct/kWh for an example of mean values over 20 years of operation.

Most of the maintenances cost are expressed as a fixed amount per year for the regular service of the turbines, but some people prefer to use a fixed amount per kWh of output in their calculations, for example around 0.01 €/kWh. The reasoning behind this method is that tear and wear on the turbine generally increases with increasing production.

In addition to maintenance, another important cost element is insurance costs. Insurance of wind turbines is required to secure the loss of their heavy investment (fully or partially as damage of parts) due to any unforeseen damage, such as lightning or hurricanes, or any other factor.

Other than the economies of scale which vary with the size of the turbine, as mentioned above, there may be economies of scale in the operation of wind parks rather than individual turbines. These economies are related to the semi-annual maintenance visits, surveillance and administration, etc.

Turbine Reinvestment (Refurbishment, Major Overhauls)

Some wind turbine components are more subject to tear and wear than others. This is particularly true for rotor blades and gearboxes. Wind turbine owners who see that their turbine is close the end of their technical design lifetime may find it advantageous to increase the lifetime of the turbine by doing a major overhaul of the turbine, e.g. by replacing the rotor blades after 7–10 years. The price of a new set of rotor blades, a gearbox, or a generator is usually in the order of magnitude of 15–20% of the price of the turbine.

The Availability Factor

The figures for annual energy output assume that wind turbines are operational and ready to run all the time. In practice, however, wind turbines need servicing and regular inspection (e.g. once every six months) to ensure that they remain safe. In addition, component failures and accidents (such as lightning strikes) may disable wind turbines. The best turbine manufacturers consistently achieve availability factors up to 98%, i.e. the machines offer technical availability of up to 98% of the

Table 7.2 Percentage example for operational costs of 5.0 €ct/kWh

Maintenance, repairs, other	50%
Lease payments for land	20%
Operational management (technical and commercial)	20%
Reserves for unforeseen events	5%
Insurance	5%

time. Total energy output is generally affected by less than 2%, since wind turbines are never serviced during high winds. The availability factor is, therefore, usually ignored when doing economic calculations, since other uncertainties (e.g. wind variability) are far larger.

It is always a good idea to check the manufacturers' track record and servicing ability before buying a new wind turbine. Manufacturers in most cases offer a guarantee of availability of the machine for the initial 2–3 years. Afterwards, they prefer to enter into a sort of annual maintenance contract with the machine owner, in which they charge a fixed amount to guarantee a certain minimum availability of the system for the generation of power.

7.2 Electrical Tariffs

Different types of tariff systems are found in connection with wind energy systems due to a difference in policies which change in country/location as described below:

Feed-in tariff
The feed-in tariff scheme, as its name suggests, is based upon the principle of paying an amount to the wind energy producer that depends on the amount of electricity fed into the grid. This is done at a pre-declared rate per unit of electricity. This rate is higher than the rate of production of electricity from a conventional (using fossil or nuclear fuels) power plant. The most important aspect in a feed-in tariff system is that the grid cannot deny accepting the power generated by the wind energy system, even if it is surplus. The feed-in tariff system exists e.g. in Germany as well as in many other countries. The rates of feed-in tariff change with respect to the location within the country. To promote the use of wind energy systems by improving their economics, currently, in Germany, the feed-in-tariff rates are higher in inland locations as compared to coastal locations which are attractive for wind energy anyway.

If an energy producer owning wind turbines produces a certain amount of electricity and consumes the same amount of electricity, separate accounting is done for the electricity produced by the windmill and the electricity consumed. The feed-in tariff rate is not uniform for all renewable energy systems, e.g. in the case of photovoltaic systems, the feed-in tariff rates are sometimes higher than those of wind energy systems.

Feed-in tariff rates also exist in India, but a difference in feed-in tariff rates between coastal and non-coastal areas does not exist. Currently, the rates are the same over one particular state; they used to be highest in the state of Maharashtra and the lowest in the state of Tamilnadu.

With the Indian government planning to organize an increasing number of reverse auctions for wind energy projects, several states are rethinking whether to continue with the feed-in tariffs at all.

Many states in India, such as Karnataka Gujarat, Andhra Pradesh, have decided not to renew the power purchase agreement with feed-in-tariff mechanism for future. One of the major reasons for such major change is very low price of electricity rates obtained through auctions for Solar Photovoltaic Systems. It is understood that technology is mature enough to be competitive in open market and hence, price auctions for wind energy too can be considered as has been done in solar PV technology.

Net metering system
Net metering or net billing is a term applied to laws and programs under which a utility allows the meter of a customer with a wind turbine, to turn backward, thereby, in effect, allowing the customer to deliver any excess electricity he produces to the utility and be credited on a one-for-one basis against any electricity the utility supplies to him. For example: During a one-month period, one wind turbine in a farm house generates 300 kilowatt-hours (kWh) of electricity. Most of the electricity is generated at a time when the equipment in the farm house (refrigerator, lights, etc.) draws electricity, and the power generated is used on site. However, some is generated at night when most equipment is turned off. At the end of the month, the turbine has generated 100 kWh in excess of local needs, and that electricity has been transmitted to the utility system. Over the same one-month period, the utility system supplied to the farm house electricity of 500 kWh for use at times when the wind turbine was not or insufficiently generating. Since the meter ran backward while the 100 kWh was being transmitted to the utility system, the farm house will be billed for 400 kWh rather than 500 kWh.

The net-metering scheme can improve the economics of captive wind turbine by allowing the turbine's owner to use the excess electricity to offset utility-supplied power at the full retail rate, rather than having to sell the power to the utility at the price the utility pays for the wholesale electricity it buys or generates itself. Many utilities argue against net metering laws, saying that they are being required, in effect, to buy power from wind turbine owners at full retail rates, and are, therefore, being deprived of a profit from part of their electricity sales. However, wind energy advocates have successfully argued that what is going on is a power swap, and that it is standard practice in the utility industry for utilities to trade power among themselves without accounting for differences in the costs of generating the various kilowatt-hours involved.

Although in Europe the feed-in tariff scheme exists, the popularity of net metering is growing, worldwide. Some countries like India even offer limited flexibility in the installation of wind turbines and consumption centers. This means that they may not be exactly at the same place. In such cases, of course, the utility charges a nominal amount termed as 'wheeling charges' against use of their infrastructure/power lines for transferring power from the point of generation to the point of consumption. In such arrangements, there are, of course, some limitations such as geographical and commercial boundaries which differ from state to state.

Time dependent rates
Ideally, electricity companies are more interested in buying electricity during the periods of peak load (maximum consumption) on the electrical grid. Therefore, in some areas, power companies apply variable electricity tariffs depending on the time of day, when they buy electrical energy from private wind turbine owners. Normally, wind turbine owners receive less than the normal consumer price of electricity, since that price usually includes payment for the power company's operation and maintenance of the electrical grid, plus its profits. In locations where a feed-in tariff system exists, as in Germany, the time when power is generated becomes non-significant.

Quota system and Renewable Purchase Obligation
An additional system is that of the allocation of 'quota' of renewable energy. In this system, every producer of electricity for grid is given a 'quota', e.g. 20%. In the total electricity produced by every company, there has to be 20% share of energy coming from renewable sources such has wind and solar. In case this quota is not met, provisions of penalties are made. The European Union is discussing to adopt this system in Europe. Provisions are being developed that if a company produces more energy from renewables than its quota, e.g. 25%, it would be granted a certificate for this excess renewable energy. This certificate can be purchased by other companies that are having a lower share of renewables from the quota, e.g. 15%. By purchasing these certificates, the second company will also be considered to comply with its quota. This system is quite similar to the concept of 'emission trading' for greenhouse gases. Up to the middle of 2008, there was no consensus on bringing such a quota system to Europe. The argument was debated that with such a quota system, relatively costlier renewable energy technologies, such as solar photovoltaic systems, would not be installed at all. Companies might prefer to purchase certificates in place of installing PV systems. The concept of the quota system is still being discussed at EU level.

The concept of quota system is similar as that of 'Renewable Purchase Obligation' in India under which every state has to compulsorily consume certain fixed percentage of energy through renewable. This percentage of RPO is primarily determined by the respective State Electricity Regulatory Commission in consultation with the National Action plan on Climate change. Quite obviously, some states where renewable energy potential is more, would find it easy to achieve this target whereas other states may face difficulties, but would still like to contribute to the national action plan for reaping other benefits. For addressing this issue, 'Renewable Energy Certificate' (REC), scheme has been launched, concept of which is similar to that of carbon credit.

An REC is created when one megawatt hour (MWh) of electricity is generated from a renewable energy resource. Typically, RECs are unbundled and sold separately, from the underlying electricity generated. When purchased, the owner of REC is considered to have purchased renewable energy, which can be used for meeting the deficit of Renewable Purchase Obligation. In order to facilitate registration, issuance and trading of RECs, a central agency has been designated by the

Central Electricity Regulatory Commission of India. The renewable energy generators will have two options: either sell the renewable energy at a preferential tariff fixed by the concerned Electricity Regulatory Commission, or sell the electricity generation and environmental attributes (RECs) separately.

On choosing the second option, the environmental attributes can be exchanged in the form of REC. Price of the electricity component would be equivalent to the weighted average power purchase cost to the distribution company, including short-term power purchase but excluding renewable power purchase cost.

The REC will be exchanged only in the power exchanges approved by CERC within the band of a floor price and a forbearance (ceiling) price to be determined by CERC from time to time.

The RPO/REC scheme has been successfully applied in several countries such as US, UK, Sweden, and Australia; and many countries are adopting it now for promoting renewable energy.

Production tax incentives/Investment incentives

The Production tax incentive is a generation-based mechanism, which supports renewable energies through payment exemptions from electricity taxes, e.g. the energy tax for renewable energies, applied to all producers. Hence it is a system that affords an avoided cost on the producer side. Also the Investment incentive is a mechanism, like the name leads one to divine, to lower the costs for the investment in renewable energies so that it gets more attractive funding.

Generation Based Incentives

In many cases, it has been observed that since most of the incentives are linked with investment, there is a big attraction among investors in installing the capacity, but relatively very less or no motivation for maintaining the plant nicely and generating maximum possible amount of electricity from the installed capacity. This happens especially in the cases where, heavy tax benefits etc. are offered, and the investor is attracted due to allied reasons and not for promoting renewable energy. The purpose of Generation Based Incentive (GBI) is to shift the mechanism of payment from a installation-based to generation-based methods and for rewarding wind farms that contribute more for clean power generation. In many countries, GBI has been introduced in parallel with investment based incentives, in order to create conducive environment for investment as well as providing motivation to operate the plants efficiently. In India, the rate of GBI is presently kept as Indian Rupee 0.50 per kWh of electricity for a minimum of four years to maximum ten years period. with maximum cap of Indian Rupee 10 million per MW capacity. GBI scheme is running in parallel with existing fiscal incentive including that of accelerated depreciation in a mutually exclusive manner. However, companies can either avail accelerated depreciation or GBI, but not both.

7.3 Mechanisms to Support Funding

7.3.1 Capacity Credit

To understand the concept of the capacity credit, we look at its opposite, the power tariffs: Large electricity customers are usually charged both for the amount of energy (kWh) they use, and for the maximum amount of power (kW) they draw from the grid, i.e. customers having more connected load drawing a lot of energy very quickly have to pay more. The reason they have to pay more is that this fact obliges the power company to have a higher total generating capacity (more power plant) available. Power companies have to consider adding generating capacity whenever they give new consumers access to the grid. But with a modest number of wind turbines in the grid, wind turbines are almost like "negative consumers", since they postpone the need to install other new generating capacity. Many power companies in industrialized countries, therefore, pay a certain amount per year to the wind turbine owner as a capacity credit. The exact level of the capacity credit varies with location benefit to utility in terms of avoided installation of power plants. In some countries it is paid on the basis of a number of measurements of power output during the year. In other areas, some other formula is used. In a number of countries like Germany, no capacity credit is given, as it is assumed to be part of the energy tariff due to the fact that due to fluctuating output from wind energy system, the utility anyways has to install power plants as back up for a no wind situation.

7.3.2 Environmental Credit and Clear Development Mechanism

Many governments and power companies around the world wish to promote the use of renewable energy sources. Therefore, in developed or industrial countries, they offer a certain environmental premium to wind energy, e.g. in the form of a refund of electricity taxes etc. on top of normal rates paid for electricity delivered to the grid. In developing countries like India, due to the fact that every unit of electricity generated avoids generation of the same amount of electricity from fossil fuel based power plants, the avoided environmental emissions offer the opportunity for additional earnings through the Clean Development Mechanism (CDM) under the Kyoto Protocol. Every ton of carbon-dioxide gas saved or avoided, termed as one carbon credit, is sold to industrialized countries at rates ranging from 3–4 € to 15–20 €, depending upon the demand and supply position of credits. These additional earnings help to improve the economics of power generation, which has been one strong reason behind the rapid increase in the wind energy based installed capacity for power generation in India.

7.3.3 *Tax Benefits*

In some countries like India, several tax benefits are offered to promote wind energy systems. Major incentives or benefits are mentioned here:

(a) If the wind energy converter or its parts or components are imported from other countries, no excise duty (import tax) is applicable.
(b) Exemption/reduction in central government sales tax and general sales tax are available on the sale of renewable energy equipment in various states of India.
(c) 80% accelerated depreciation on specified renewable energy devices/systems (including wind power equipment) in the first years of installation of the projects.
(d) There is an income tax holiday on the income generated from renewable energy systems including wind energy through power generation.

As a result of these measures, several investors have installed wind energy systems in different parts of India.

7.4 Wind Energy Economics

7.4.1 *Financing of Wind Park—A Case Study for India*

The annual electricity production will vary enormously depending on the amount of wind on the turbine site. Therefore, there is not a single price for wind energy. It differs from year to year and location to location, that means wind speed and wind profiles.

Following a case study from India should give an impression about the factors which has to be considered to calculate the electricity costs from wind turbines.

It is proposed to install a 1,250 kW wind power plant at a certain location where the capacity utilization for power generation over one year is expected to be 35%. Below are the major cost elements in Indian rupees (Rs.):

Currency (June 2017)	100 Rs. = 1.40 €; 1 € = 72 Rs.
	100 Rs. = 1.540 US$; 1 US$ = 65 Rs.
Cost of land	Rs. 1 million
Cost of wind energy converter	Rs. 60 million
Erection and commissioning charges	Rs. 6 million
Sub station charges	Rs. 4.0 million

Operation and maintenance charges: Free during first year, second year Rs. 1.1 million, for subsequent years with an escalation of 5% per year.

Other economic parameters include a total generation as 3.85 million kWh electricity (at 35% capacity utilization), and losses as 7%.

Tariff for selling the electricity is Rs. 5.5/kWh in the first year, with a price escalation of 2% per year.

Wheeling charges: 5% of tariff.

Life cycle term of the project is 20 years.

The investment requirement, power generation and sale of power can be summarized as in the Table 7.3.

Table 7.4 shows a detailed cash flow analysis of the project under various heads:

It can be seen from the above table that the cumulative cash accrual becomes equal to the initial investment of Rs. 71 million at the end of the fourth year. This period is known as the simple payback period of investment. Beyond this period and up to the end of project life (20 years in this case), every cash accrual is net profit from the project. If the financial discounting rate is significant and the payback period is longer, it is recommended to include the time based value of money for estimating the payback period.

Table 7.3 Assumptions for the case study of financial analysis

Investment requirement	
Cost of land (million Indian rupees)	1
Cost of wind machine (million Indian rupees)	60
Erection & commissioning charges (million Indian rupees)	6
Sub-station charges (million Indian rupees)	4
Total cost (million Indian rupees)	71
Running cost requirement	
O & M (million Rs.)	1.1
Escalation	5%
Free O & M (years)	1
Depreciation rate	80%
Insurance (million Rs./year)	0.15
Tariff (Rs./kWh)	5.5
Escalation in tariff per year	2%
Wheeling charges (% of tariff)	5%
Generation (million kWh)	3.85
Losses	7%
Sale of power	
Tariff (Rs./kWh)	5.5
Wheeling charges (5% of tariff)	0.28
Actual tariff (tariff—Wheeling charges)	5.23
Generation (million kWh)	3.85
Losses (7% of Generation)	0.27
Actual generation (Generation—Losses) kWh/year	3.58

Table 7.4 Cash flow analysis of wind energy project

Year	1	2	3	4	5	6–20
Receipt (million Rs.)						
Power rate (Rs./kWh)	5.23	5.28	5.33	5.38	5.44	Increasing as per rate of tariff
Receipt (sale of power)	18.71	18.90	19.08	19.27	19.47	Increasing from 6th year to 20th year
Total receipts (A)	**18.71**	**18.90**	**19.08**	**19.27**	**19.47**	**Increasing from 6th year to 20th year**
Expenses (million Rs.)						
C & M (free for 1st year)	0	1.1	1.155	1.21	1.27	Increasing from 6th year to 20th year
Insurance	0.15	0.15	0.15	0.15	0.15	0.15
Total expenses (B)	**0.15**	**1.25**	**1.31**	**1.36**	**1.42**	**Increasing from 6th year to 20th year**
Cumulative cash accrual (A–B)	**18.56**	**36.20**	**53.98**	**71.89**	**89.94**	**Increasing up to 20th year (project life)**

7.4.2 Financing of a Wind Park—Description of a Case in Germany

Since May 2017, the Federal Network Agency will issue tenders for the determination of the support of wind energy installations on land. The calculated value to be used serves as the basis for the calculation of the amount of the payment claim (market premium). Within the framework of the announcement, a power to be installed in kilowatts is issued for one or more approved plants at a bid price for the electricity generated therein. The bids must therefore refer to a specific value to be applied in cents per kilowatt-hour (bid value) for the electricity generated in the plants and to a plant output in kilowatt (bid amount). Although, the bids with the lowest bids receive the addition, until the volume of the bidding is reached. The amount of the payments will be determined for all new sites, which have been put into operation since January 2017 with an installed capacity of more than 750 kW, by means of invitations to tender. Each year, three to four bids are held, with a funding period of 20 years. In addition, it is stipulated that the expansion of new wind onshore does not exceed 2.9 GW/year, including the replacement of old stations.

To raise the equity capital a limited partnership (L.P.) is founded. This L.P. incurs the sale of the interests and covers the banker's guarantee. It is also responsible for the construction, the instigation and the disposal of the produced electric power. The limited partner's individual deposit constitutes the equity capital and starts from several thousand euros per partner. The bank allocates the dept capital in form of a loan. This loan could be for example a loan of 11 million €, with a duration of 15 years and an interest loan of about 6%. As a chattel mortgage the wind park, the feed-in tariffs and the liquidity reserves are transferred to the bank until the loan is paid off. The money for the bank has priority over the payout of dividends for the limited partners.

The funding of an offshore wind park requires an investment volume of at least 200 million € and more. Therefore, a limited partnership cannot be established. Electricity companies are taking over this part beside some private investors.

The first offshore wind parks will receive the duple amount of money for their produced electricity than the on shore wind parks. Therefore it is possible to operate these offshore wind parks.

7.5 Wind Turbines after Operational Life

The amount of material used in wind turbines increase with the size and the number of wind turbines. After the operational life of the equipment it must be demolished and as far as possible recycled. Some components of wind energy converter, such as generator or gear box do not cause much problems due to their material such as copper, steel, being reusable as well as recyclable. The major concern about disposal is related to blades since it is not easy to recycle them since they are made of plastic, glass fibers or carbon fibers.

Presently, there are three possible modes for dismantled wind turbine blades: landfill, incineration or recycling. The first option, landfill, has very limited possibility due to the fact that many countries have already started imposing a ban on landfills with glass-fibre reinforced plastic due to high organic content, that is extensively used in making of blades.

The next possibility is incineration, in which electricity can be generated by using the heat released. Due to the presence of large percentage of inorganic composites in blades, large residue would be remaining after incineration leading to other problem of ash dumping. The inorganic material may also lead to the emission of hazardous flue gasses.

The last alternative left is recycling—either material recycling, or product recycling in the form of re-powering where old turbines are replaced by newer, more efficient ones. Presently, there are very few units where recycling of wind turbine blades is taking place.

Therefore industry has to start to develop solutions how to get rid the waste in ecological acceptable waste treatment produces. Research thrust is also required to develop blade materials that are easy for reuse and recycle. Some aspects of this stage and their impact on energy and environment are discussed through a case study in the next chapter of Life cycle Analysis of wind energy converters.

Literatures

Annual report 2016–17. Indian Renewable Energy Development Corporation, http://www.ireda.in/incentives.asp
Deutsche Windguard, Kostensituation der Windenergie an Land in Deutschland, Projektnummer VW 15110, Dec. 2015, http://www.windguard.de
National Institute for Wind Energy, http://www.niwe.res.in

Chapter 8
Life Cycle Assessment of a Wind Farm

Life Cycle Assessment (LCA) is an important tool for industry and policy makers, used to determine the actual emissions of a product or technology throughout its whole life cycle. In case of energy production systems or power plants, analysis of energy required to produce the materials and processes; emissions resulting from various processes for materials production and processes resulting into their Cumulated Energy Demand (CED) and Global Warming Potential (GWP) become important parameters when making decisions on further research, development and deployment of any technology. The method of carrying out such analysis is explained in this chapter through a case study.

8.1 Basic Targets and Approach

The four different steps of a life cycle assessment are defined in the international standards DIN EN ISO 14040 and DIN EN ISO 14044:

The **Goal and Scope Definition** includes the description of the balanced object, the system boundaries and the assumptions made.

During the **Inventory Analysis** all material, energy and emission flows (input and outputs) are investigated and lead to the life cycle inventory analysis result.

Within the **Impact Assessment**, the impact categories as well as their indicators are defined. The inputs and outputs are classified by these impact categories. Using characterization factors, the impact category indicators are calculated.

Finally, in the **Interpretation** step, the results of the life cycle assessment are evaluated to reach conclusions. Additionally, an optional sensitivity analysis can investigate responsive parameters and their influence on the results.

Three different methodological approaches exist to calculate the above mentioned indicators:

H.-J. Wagner and J. Mathur, *Introduction to Wind Energy Systems*, Green Energy and Technology, https://doi.org/10.1007/978-3-319-68804-6_8

- the energetic input output analysis,
- the process chain analysis and
- the material balance analysis.

When using the input output analysis, there is a high possibility of receiving inaccurate results. A process chain analysis depends on detailed process data and is an extremely time consuming balancing process. A good compromise between a high accuracy of the result and an acceptable expenditure of time can be reached by a material balance analysis. Another advantage is that data concerning mass, material and type of production are sufficient to calculate the indicators.

8.2 Case Study of Alpha Ventus

Results and characteristics of wind energy LCA can be best described using an example. Therefore in the following chapters the carried out LCA of the offshore wind farm alpha ventus will show the different steps and important conclusions of results of LCA of a wind energy system.

8.2.1 Goal and Scope Definition

The object of investigation is the first German offshore wind farm alpha ventus beyond the territorial waters (twelve mile zone) in the German exclusive economic zone. Alpha ventus is a test field in the North Sea for 5 MW wind energy converters in a water depth of 30 m (Fig. 8.1). The field consists of six converters of the type 5 M (manufacturer REpower) that are installed on Tripod foundations and six converters of the type M5000 (manufacturer Multibrid) that are built on Jacket foundations (Fig. 8.2). All converters are connected by a 16 km submarine cable at 30 kV. An offshore transformer station steps up the voltage for the 110 kV submarine cable. The electricity is transported over 60 km to an onshore transformer station, which supplies the German high-voltage grid. Several indicators are taken into consideration and investigate the resource conserving and environmentally friendly way of power generation in comparison to the German grid mix. The functional unit was defined as the supply with 1 kWh current generated by alpha ventus at the high-voltage grid at the onshore transformer station at the coast.

8.2.2 Inventory Analysis

In a first step, the inventory—in this case all the component parts and structural elements of the wind farm—was listed. In a second step, the inventory analysis

Fig. 8.1 Wind farm alpha ventus (photo: © DOTI 2009)

Fig. 8.2 Installation of a wind converter of the wind farm alpha ventus (photo: © DOTI 2009)

Table 8.1 Overview of system boundaries

Description	Technical lifetime	Full load hours[a]	Maintenance assignment[b]	Number of WEC
alpha ventus	Foundation and Submarine Cable 20 years	3,900 h/a	10 Helicopter and 15 Shipping Services per year and WEC	12

[a]Including down time and losses during transmission to on-shore transformer station
[b]Maintenance for each wind energy converter (WEC) over the technical lifetime: replacement of 0.5 gearboxes and 1.25 rotor blades

results for the manufacturing phase, the use phase, as well as the disposal phase was built. The main part of the production phase was the calculation of the raw material and their respective masses (under consideration of common proportions of recycled material) and the subsequent processes. Additionally, the transport of the components between the different production facilities and to the building site was included, as well as the assembling of the components (wind energy converters, transformer stations and submarine cables). In conjunction with the operating company, the maintenance and maintenance assignments for the use phase for each wind energy converter were assumed (Table 8.1). Part of the disposal phase was the dismantling and the return transports of the structural elements. The inventory analysis result was built in cooperation with respective manufacturers and the operator of the wind farm.

8.2.3 Impact Assessment

In the next step of the LCA, specific energy expenditures and produced emissions were linked to the inventory analysis result. The database of the Swiss Center for Life Cycle Inventories Ecoinvent lists over 4,000 datasets for specific energy expenditures and produced emissions for raw material. Furthermore, data for subsequent processing, like forging or welding, can be withdrawn. With the help of the balancing software GaBi and the implemented Ecoinvent database, the linking process could be performed and the offshore wind farm analyzed holistically.

Cumulated energy demand
As the routing indicator the cumulated (primary) energy demand (CED) was chosen, which sums up every energetic input over the life cycle of a product evaluated on a primary energy basis. It is measured in TJ primary energy-equivalent (PE-Eq.). It allows an energetic comparison to other power generating systems and their respective CED.

Global warming potential
Due to the current public focus on the reduction of greenhouse gases, the global warming potential or greenhouse warming potential cumulates every carbon

dioxide equivalent emission. Hence, it describes the anthropogenic global warming effect, which leads to the increase of the global average temperature. It is measured in kg CO_2-Equivalent.

Additional indicators
In LCAs further indicators, such as the eutrophication potential (EP), the human toxicity potential (HTP), the photochemical ozone creation potential (POCP) and the acidification potential (AP), can be investigated. Depending on the Scope Definition of the LCA, these categories can deliver important results. However, they will not be part of the example described here.

8.2.4 Interpretation and Results

The results are organized to give an overview of the CED, itemized with respect to the life cycle phases of the wind farm. Table 8.1 shows all assumptions with respect to the technical lifetime of the foundation and the submarine cable, the full load hours, the maintenance and the maintenance assignments. For the wind farm, Fig. 8.3 displays the total material balance itemizing different substances. All together about 29,000 tons of material are installed. Of major significance is the proportion of the ferrous metal (>73%). Especially the foundations of the offshore installations, which contribute more than half of the ferrous metal, have a considerable influence. Non-ferrous metal has a proportion of about 8%. In the first approximation, the submarine cable adds up to 80% of the non-ferrous metal.

CED and GWP results respecting the life cycle phases
In total, about 2,300 TJ PE-Equivalent have to be expended for the entire life cycle of alpha ventus (Fig. 8.4). A proportion of nearly 80% is attributed to the production phase, followed by the use phase (about 20%). The disposal phase contains the deconstruction and has a minor influence compared to the other life cycle phases.

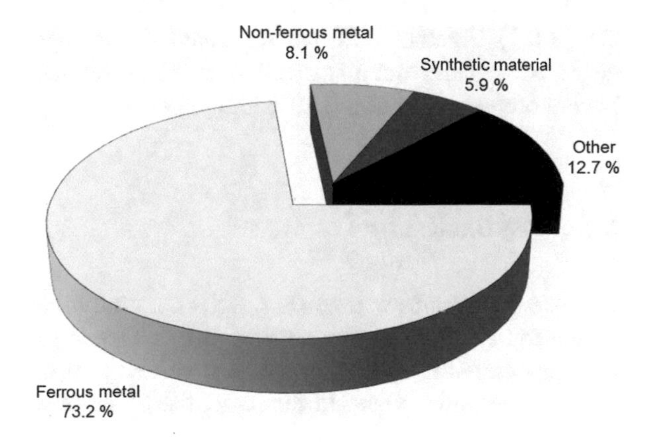

Fig. 8.3 Material balance of the wind farm alpha ventus (including the grid connection)

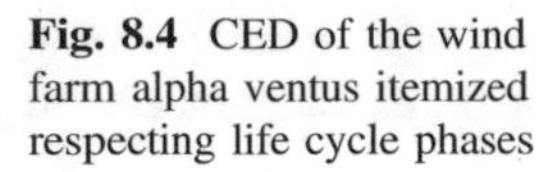

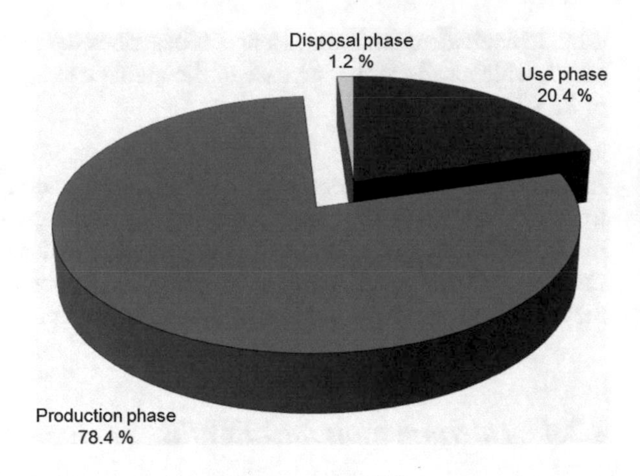

Fig. 8.4 CED of the wind farm alpha ventus itemized respecting life cycle phases

Concerning the GWP, about 149,000 t CO_2-Equivalent are expended. The proportions of the different life cycle phases nearly equal the respective one of the CED (Fig. 8.4). This is on account of the high proportion of the CO_2 emissions (about 90%) in the GWP, which are mainly caused by the expenditure of fossil energy. Thus, they are related to the CED. Therefore the following remarks can also give conclusions to those of the GWP.

Production phase

Different structural components have different influences on the total CED results. The main influence of a single component has the 110 kV submarine cable. Its production contributes about 12% to the total CED. Due to the high impact of the submarine cable, the distance between the wind farm and the shore is also of major importance. The offshore and onshore transformer stations sum up to 6%. Each of the 12 converter contributes about 2% to the CED, the 110 kV submarine cable about 12% in total. All twelve foundations add up to an amount of 35% to the total CED.

Use phase

Investigating the use phase, the calculations show that about 70% of the 470 TJ PE-Equivalent for the use phase (Fig. 8.4) are expended for shipping services (Table 8.1). Another 12% are expended for transports by helicopters. Changing gear boxes, blades and oil as well as the transport and assembling of the spare parts have a proportion of about 18% altogether.

8.3 Payback Time

The energetic payback time (Fig. 8.5) is an important key figure to determine the sustainability of power plants using renewable energies. It describes the time frame to compensate the energetic expenses, valued as primary energy, for the entire life cycle of the power plant. In this case, the offshore wind farm generates electricity

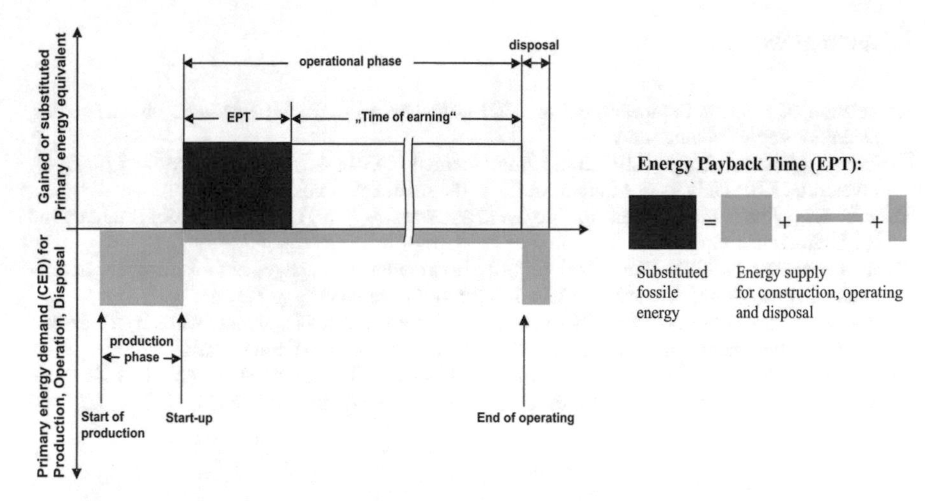

Fig. 8.5 Energy payback time

Table 8.2 Energetic and greenhouse gas payback period

Description	Generated electricity[a] over 20 years	Energetic payback period[b]	Greenhouse gas payback period[c]
alpha ventus (60 MW)	4,680 GWh	8.8 Month	9.1 Month

[a]Unweighted = secondary energy
[b]Energetic supply factor for the German electricity mix at the high-voltage grid: 3.007 kWh PE-Eq./kWhel
[c]Greenhouse gas supply factor for the German electricity mix at the high-voltage grid: 0.665 kg CO_2-Eq./kWhel

without any additional energetic input to compensate the energetic expenses for production, use and disposal. To value the generated electricity as primary energy, Germany's electricity mix was chosen as the comparing system. The calculations are based on the computational method of VDI 4661.

To investigate the influence on the climate change, the greenhouse gas payback period was calculated, too. This key figure describes the time for the compensation of the carbon dioxide equivalent emissions during the entire life cycle. Again, as the comparing system, the German electricity mix was chosen (Table 8.2).

The results show that the energetic input for the wind farm is amortized after less than 10 months. The greenhouse gas payback period is about a half month longer than the energetic payback period. The results show a strong correlation between each other. Due to the effect that the CO_2 emissions are mainly based on the energetic expenditures, the high proportion of about 90% of the CO_2 emissions in the emitted greenhouse gases lead to this correlation. Therefore, within less than one year the energetic expenditure as well as the greenhouse gas emission of the entire life cycle of alpha ventus is amortized.

Literatures

Association of German Engineers, editor. VDI-guideline 4661 (2003) Energetic characteristics, Definition-terms-methodology

Ecoinvent database, Version 2.01 (2007) Implemented in Gabi 4.3, Swiss Centre for Life Cycle Inventorie, LBP University Stuttgart and PE International GmbH

GaBi V. 4.3 software and database for life cycle assessments (2007) LBP University Stuttgart and PE International GmbH

German Standard DIN-EN-ISO-14040 (2006) Environmental management—Life cycle assessment—Principles and framework (ISO 14040:2006). BeuthVerlag, Berlin

German Standard DIN-EN-ISO-14044 (2006) Environmental management—Life cycle assessment—Requirements and guidelines (ISO 14044:2006). Beuth Verlag, Berlin

Wagner H-J, Baack C, Eickelkamp T, Epe A, Lohmann J, Troy S et al. (2011) Life cycle assessment of the offshore wind farm alpha ventus. Energy 36: 2459–2464, ISSN 0360-5442

Chapter 9
Outlook

More than 50 countries around the world have an organized set up in the field of wind power. With China emerging as world leader, surpassing US, large progress in wind power has been witnessed in the countries of the European Union especially Germany. Developing countries like India, Brazil, have also added significantly large capacities over past few years. New wind energy markets are also emerging in Africa, with South Africa, Morocco and Kenya giving a special attention to their renewable energy programs. In the past few years, a new frontier for wind power development has also been established in the sea. With offshore wind parks beginning to make a contribution, the possibilities of using wind energy systems worldwide are likely to increase many times in the future. Establishing wind energy projects in the sea has opened up new demands, including the need for stronger foundations, long underwater cables and larger individual turbines, but offshore wind parks are contributing an increasing proportion of global capacity.

With its increase in the global market, wind power has seen a fall in cost. A modern wind turbine produces more electricity at less cost per unit (kWh) than its equivalent ten years ago. At good locations wind may compete with, and even beat, the cost of both coal and gas-fired power, if the future carbon dioxide certificate costs are taken into account. If the "external costs" associated with the pollution and health effects resulting from fossil fuel and nuclear generation are fully taken into account, wind power gains an additional advantage.

Studies of different institutions around the world have confirmed that a lack of wind is not likely to be a limiting factor on global wind power development during next decade because the potentials are not used until yet. With the expansion of the wind industry, large quantities of wind powered electricity will need to be integrated into the grid network. Because of the variability of the wind, control methods have to be established for dealing with variations in demand and supply. Their installation could help to handle this issue. Also, the extension of the electrical grids and the building of new load-flexible conventional power stations is going on the way.

H.-J. Wagner and J. Mathur, *Introduction to Wind Energy Systems*, Green Energy and Technology, https://doi.org/10.1007/978-3-319-68804-6_9

The hope is also that energy storages, like batteries, become available with good quality and economically.

On one side, economic instabilities worldwide, changing growth rates, natural calamities may be hindrances in the growth of wind energy sector, however, other factors, such an increasing concern for environment and natural resources, price of oil, research break-through and economy of scale are expected to promote this sector.

The worldwide installed wind power capacity was already 500 GW at the end of the year 2016.

It has been further estimated by the Global Wind Energy Council that wind power could supply one third of the world's electricity by 2050. This is a big dream and challenge not only for the wind energy sector, but for the entire electricity supply system. Let's try to realise it.

Literature

Global wind energy outlook 2014–15 Greenpeace International (www.greenpeace.org); Global wind energy council (GWEC) (www.gwec.net)

About the Book

Authors have tried to strike a balance between a short book chapter and a very detailed book for subject experts. There were three prime reasons behind doing so: first, the field is quite interdisciplinary and requires simplified presentation for a person from non-parent discipline. Second reason for this short-version of a full book is that both the authors have seen students and technically oriented people, searching for this type of book on wind energy. Third reason and motivations was considering engineers who are starting their career in wind industry. This book is targeted to present a good starting background to such professionals.

H.-J. Wagner and J. Mathur, *Introduction to Wind Energy Systems*, Green Energy and Technology, https://doi.org/10.1007/978-3-319-68804-6

Glossary

Alternating Current (AC) An electric current that reverses its direction at regularly recurring intervals, usually 50 or 60 times per second. Today's grids are operated by Alternating Current.

Direct Current (DC) An electric current with constant direction. I can be used on high voltage level to transport bulk electricity with less losses.

Joule (J) A standard international unit of energy; 1,055 J are equal to 1 BTU.

Watt (W) The rate of energy transfer equivalent to 1 A under an electrical pressure of 1 V. 1 W equals 1/746 horsepower, or 1 J/s. It is the product of voltage and current (Amperage).

Watthour (Wh) The electrical energy unit of measure equal to 1 W of power supplied to, or taken from, an electric circuit steadily for 1 h.

Kilowatt (kW) A standard unit of electrical power equal to 1,000 W, or to the energy consumption at a rate of 1,000 joules per second.

Kilowatt-Hour (kWh) 1,000 W acting over a period of 1 h. The kWh is a unit of energy. 1 kWh = 3,600 kJ.

Megawatt (MW) 1,000 kW, or 1 million watts, standard measure of electric power plant generating capacity.

Megawatt-Hour (MWh) 1,000 kWh or 1 million watt-hours.

Gigawatt (GW) A unit of power equal to 1 billion Watts, 1 million kilowatts, or 1,000 megawatts.

H.-J. Wagner and J. Mathur, *Introduction to Wind Energy Systems*, Green Energy and Technology, https://doi.org/10.1007/978-3-319-68804-6

Index

H.-J. Wagner and J. Mathur, *Introduction to Wind Energy Systems*, Green Energy and Technology, https://doi.org/10.1007/978-3-319-68804-6

W

Y

Printed in the United States
By Bookmasters